International Conference on Advanced Functional Materials and Composites (ICAFMC 2018)

AA004473

MATEC Web of Conferences Volume 242 (2018)

Barcelona, Spain
5 - 6 September 2018

Editor:

Ana I. Almerich-Chulia

ISBN: 978-1-5108-7596-8

Printed from e-media with permission by:

Curran Associates, Inc.
57 Morehouse Lane
Red Hook, NY 12571

Some format issues inherent in the e-media version may also appear in this print version.

Printed by Curran Associates, Inc. (2019)

For additional information, please contact EDP Sciences – Web of Conferences
at the address below.

EDP Sciences – Web of Conferences
17, Avenue du Hoggar
Parc d'Activité de Courtabœuf
BP 112
F-91944 Les Ulis Cedex A
France

Phone: +33 (0) 1 69 18 75 75
Fax: +33 (0) 1 69 28 84 91

contact-edps@webofconferences.org

Additional copies of this publication are available from:

Curran Associates, Inc.
57 Morehouse Lane
Red Hook, NY 12571 USA
Phone: 845-758-0400
Fax: 845-758-2633
Email: curran@proceedings.com
Web: www.proceedings.com

TABLE OF CONTENTS

Preface

Dear Distinguished Authors and Guests,

The 2018 International Conference on Advanced functional Materials and Composites (ICAFMC2018) was taken place in Barcelona, Spain on September 2018.

The development of functional materials and their composites is a domain that has been growing very fast in the last few years. ICAFMC2018 covers all relevant aspects of composites research included in but not limited to the overall professional theme areas from material processing, development, characterization to application functional materials and their composites. It provides a high-standard international forum to introduce, to exchange and to discuss recent advances novel and practical techniques or application in the field of material engineering and application.

As an academic conference, this event aims to bring together leading academic scientists, researchers, scholars and students around the world to exchange and share their experience and research fruits with professionals and industrial engineers.

All papers were accepted through a peer-review process by the conference program committee and international reviewers. We hope that the conference results constituted significant contribution to the knowledge in these up to date scientific field.

We would like to thank the program chairs, organization staff, and the members of the program committee for their work. Thanks also go to all those who have contributed to the success of ICAFMC 2018. We hope that all participants and other interested readers benefit scientifically from the proceedings and find it stimulating in the Process.

With my warmest regards,

Guest Editor Conference
Dr. Ana Almerich-Chulia
Universitat Politècnica de València, Spain
October 2018

ICAFMC2018 Committee

Statement of Peer review

In submitting conference proceedings to *Web of Conferences*, the editors of the proceedings certify to the Publisher that

1. They adhere to its **Policy on Publishing Integrity** in order to safeguard good scientific practice in publishing.
2. All articles have been subjected to peer review administered by the proceedings editors.
3. Reviews have been conducted by expert referees, who have been requested to provide unbiased and constructive comments aimed, whenever possible, at improving the work.
4. Proceedings editors have taken all reasonable steps to ensure the quality of the materials they publish and their decision to accept or reject a paper for publication has been based only on the merits of the work and the relevance to the journal.

Title, date and place of the conference

International Conference on Advanced Functional Materials and Composites (ICAFMC 2018)
September 5-6, 2018 in Barcelona, Spain

Proceedings editor(s):

Ana I. Almerich-Chulia, Universitat Politècnica de València, Spain
(analchu@mes.upv.es)

Date and editor's signature

5th November 2018

Technology and electrophysical properties of Mn^{4+}, Sb^{3+}, Dy^{3+} and W^{6+} -doped $Pb(Zr_{0.49}Ti_{0.51})O_3$ ceramics

Dariusz Bochenek[1,a], Ewa Ćwikiel[1], Przemysław Niemiec[1] and Tomasz Goryczka[2]

[1] *University of Silesia in Katowice, Faculty of Computer Science and Material Science, Institute of Technology and Mechatronics 12, Żytnia St., 41–200, Sosnowiec, Poland*
[2] *University of Silesia in Katowice, Faculty of Computer Science and Material Science, Institute of Material Science, 1a, 75 Pułku Piechoty St., 41–500 Chorzów, Poland*

Abstract. In paper three multicomponent PZT-type ceramics doped by Mn^{4+}, Sb^{3+}, Dy^{3+}, W^{6+} have been prepared by the conventional mixed oxide method. The multicomponent PZT-type ceramic powders were synthetized by calcining route, while densification was carried out by pressureless sintering method. The paper presents XRD, SEM, EDS, dielectric and DC electric conductivity measurements of the multicomponent PZT-type ceramic samples. Conducted tests indicate obtained multicomponent PZT-type exhibit good dielectric properties giving the possibility for their use as actuators and piezoelectric transducers in modern micromechatronic and microelectronic applications.

1 Introduction

Ceramic materials prepared on the basis of lead zirconate titanate $Pb(Zr_{1-x}Ti_x)O_3$ (PZT-type ceramics in short) are one of the most studied ferroelectric materials, due to its extremely wide field of electronic 'smart' applications including sensors, actuators, transducers, transformers, electric resonators, speakers, nonvolatile or computer memories, display, MEMS devices etc. [1-5]. The properties of PZT-type ceramics depend on the ratio of Zr^{4+}/Ti^{4+} and on the admixtures introduced to the base composition [6-10, 11]. In the room temperature pure PZT has ferroelectric properties (so it possesses also piezo and pyroelectric properties) for: $0.042<x<0.380$ (rhombohedral phase $R3c$), $0.380<x<0.470$ (rhombohedral phase $R3m$) as well as $0.480<x<1.000$ (tetragonal phase $P4mm$) [5, 12]. PZT solid solutions from the area of $0.47<x<0.48$ is called morphotropic phase boundary region (MPB). PZT ceramic materials with compositions near to the MPB reveal excellent electromechanical properties [13-15].

Dopants are added to PZT material in order to enhance the electrophysical properties. Additives alter the defect structure of the PZT lattice and give the opportunity to tailor the electrophysical properties in a wide range. Is common practice in planning new compositions to use more than one type of dopants to the given composition in order to achieve the appropriate set of properties [16]. Appropriate choice of a type and a quantity of dopant ions is important.

In the research the multicomponent PZT-type ceramics doped by manganese Mn^{4+}, antimony Sb^{3+}, dysprosium Dy^{3+} and tungsten W^{6+} were obtained and investigated.

[a] dariusz.bochenek@us.edu.pl

The three chemical compositions:
$Pb(Zr_{0.49}Ti_{0.51})_{0.94}Mn_{0.021}Sb_{0.016}Dy_{0.012}W_{0.012}O_3$ (PZT-1),
$Pb(Zr_{0.49}Ti_{0.51})_{0.94}Mn_{0.021}Sb_{0.016}Dy_{0.010}W_{0.013}O_3$ (PZT-2),
$Pb(Zr_{0.49}Ti_{0.51})_{0.94}Mn_{0.021}Sb_{0.016}Dy_{0.008}W_{0.014}O_3$ (PZT-3),
were designed and obtained, at following technological conditions: T_s= 1170°C/t_s=2 h.

The aim of this study was obtained and examined Mn^{4+}, Sb^{3+}, Dy^{3+} and W^{6+} - doped $Pb(Zr_{0.49}Ti_{0.51})O_3$ multicomponent ceramic materials for modern microelectronic and micromechatronic applications.

2 Experimental details

2.1 Technology and sample preparation

Ceramic samples of the multicomponent PZT-type materials were prepared by the conventional mixed oxide method. The starting powders PbO (99.99 %, POCH), ZrO_2 (99.00 %, Merck), TiO_2 (99.99 %, Merck), MnO_2 (90 %, Merck), Sb_2O_3 (99.995 %, Aldrich), Dy_2O_3 (99.9%, Aldrich), and WO_3 (99.9 %, Fluka) were milled for 24 h in ethanol using ZrO_2 balls. After drying the mixtures of powders were calcined in air at following conditions T_{calc}= 850°C for t_{calc}= 4 h, at heating rate of 150°C/h. An additional 5.0 wt.% PbO was added to compensate for its evaporation during calcination and densification. Densification (sintering) was carried out by the conventional method (pressureless sintering method). Disks 10 mm in diameter and 1 mm in thickness were pressed and next were sintered in closed alumina crucibles at following sintering conditions: T_s= 1170°C / t_s=2 h (a heating rate of 150°C/h). The final steps of technology were grinding, polishing, removing mechanical stresses (annealing was performed at temperature at following conditions: 700°C/15 min.) and for electrical testing the surfaces of the samples were covered with a silver electrode.

2.2 Characterization

X-ray measurements were carried out at room temperature using X'Pert-Pro PW3040/60 diffractometer. The copper radiations $CuK_{\alpha 1}/_{\alpha 2}$ were used. The X-ray diffraction patterns were stored at range of 2θ from 10° to 130° in step-scan mode: 0.05 degrees and 4 s/step.

The chemical composition and microscope observation of the samples were done using JEOL JSM-7100 TTL LV Field Emission Scanning Electron Microscope equipped with Energy Dispersive Spectrometer (EDS).

Temperature studies of the dielectric properties were made using a QuadTech 1920 Precision LCR Meter at temperature range from 20°C to 450°C (heating cycle, heating rate of 1.0 deg./min., frequency range of the measurement field from 20 Hz to 1 MHz). Measurements of DC electric conductivity were made using a 6517B Keithley electrometer in the temperature range from 20°C to 450°C (for 5 V).

3 Results and discussion

3.1. X-ray diffraction analysis

Fig. 1 shows comparison of the X-ray diffraction patterns measured for the multicomponent PZT-type ceramics at room temperature (T_r). The first of all, X-diffraction lines, presented in the patterns, were identified as a one belonging to the perovskite type of structure with tetragonal crystal lattice.

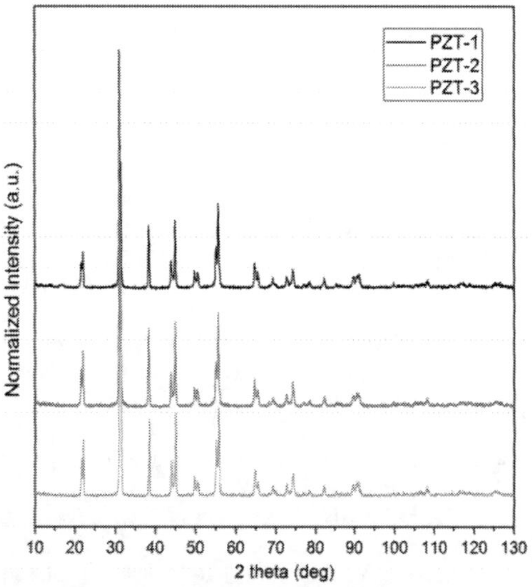

Figure 1. X-ray diffraction patterns of the multicomponent PZT-type ceramics.

All the PZT-type materials were a single phase without the presence of substrate phases or other equilibrium one. In order to determine the lattice parameters, the X-ray diffraction patterns were fitted using the Rietveld method [17]. Data for the starting model were received from ICDD data base for polymorphic phase: $Ba_{0.8}Pb_{0.2}TiO_3$ (card no. 074-2492). Symmetry of the crystal lattice can be described using space group *P4mm*. Results of determined lattice parameters were shown in Table 1.

Table 1. Values of the parameters of the multicomponent PZT-type samples (dielectric studies were carried out for ν=1 kHz).

	PZT-1	PZT-2	PZT-3
ρ (g/cm^3)	6.96	6.81	7.05
a_0 (Å)	4.0343	4.0363	4.0371
b_0 (Å)	4.0343	4.0363	4.0371
c_0 (Å)	4.1225	4.1249	4.1258
V_0 (Å^3)	67.09	67.20	67.24
T_C (°C)	284	290	296
ε_r at T_r	447	848	680
ε_{max} at T_C	8486	15990	12386
tanδ at T_r	0.003	0.002	0.004
tanδ at T_C	0.122	0.089	0.104
ρ_{DC} at T_r (Ωm)	3.4 x10^9	4.7 x10^8	5.3 x10^7

3.2. Microstructural and EDS tests

Fig. 2 shows the microstructural SEM images of the PZT-type compositions which prove that the technological process has been properly carried out and correctly selected technological conditions.

On the SEM microstructural image, the PZT-1 ceramic sample (with the largest amount of dysprosium and the smallest amount of tungsten) ceramic grains exhibit a correct angular shape witch large size (Fig. 2a). The grain boundaries of the PZT-1 ceramics have regular shapes. The fracture of the PZT-1 ceramic sample occurs mainly along the grain boundary, and in a small percentage the fracture is also carried out through the grain.

In the case of the PZT-2 ceramic sample (with the intermediate amount of tungsten and dysprosium dopant) an increase in the heterogeneity of the microstructure grains is observed (Fig. 2b). The microstructure of the PZT-2 ceramics also has grains with non-sharp edges and irregular shape of grain boundaries and the fracture occurs both on the grain boundaries and through the grains. In case of the PZT-3 ceramics the microstructure appearance is similar to the microstructure of the PZT-1 ceramics.

The PZT-type ceramic samples were subjected to composition homogeneity tests by spectroscopy with EDS energy dispersion (Fig. 3) based on point and surface analysis. The EDS tests confirmed the qualitative composition of the obtained samples without the presence of foreign elements. Summarized in the Table 2 theoretical (Theor.) and calculated (Exptl.) the percentage

https://doi.org/10.1051/matecconf/201824201001

of the individual components of the PZT-type ceramics (the average of ten randomly chosen areas from the specimen surface). In the case all samples, zirconium, titanium and introduced admixtures (i.e. manganese, antimony and tungsten) deficiency are observed compared to theoretical calculations. In the same time an excess of lead and dysprosium are observed. All deviations from the initial composition are within the acceptable range.

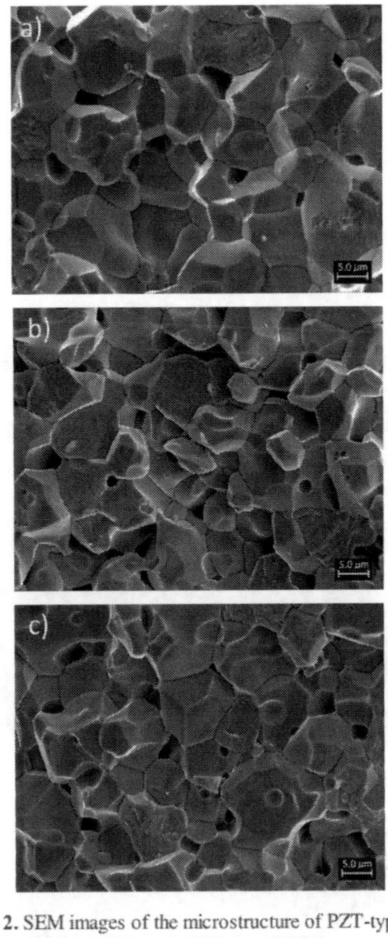

Figure 2. SEM images of the microstructure of PZT-type sample fractures: a) PZT-1, b) PZT-2, c) PZT-3.

Table 2. Theoretical and experimental percentages of elements of PZT-type ceramics.

	PZT-1		**PZT-2**		**PZT-3**	
	Theor. (%)	Exptl. (%)	Theor. (%)	Exptl. (%)	Theor. (%)	Exptl. (%)
PbO	66.60	71.66	66.63	70.92	66.67	71.88
ZrO_2	18.02	14.52	18.03	14.39	18.03	14.51
TiO_2	12.16	10.53	12.17	11.69	12.17	11.08
MnO_2	0.58	0.44	0.58	0.47	0.58	0.49
Sb_2O_3	0.78	0.70	0.85	0.64	0.93	0.60
Dy_2O_3	0.83	1.4	0.71	1.25	0.59	0.76
WO_3	1.03	0.74	1.03	0.64	1.03	0.69

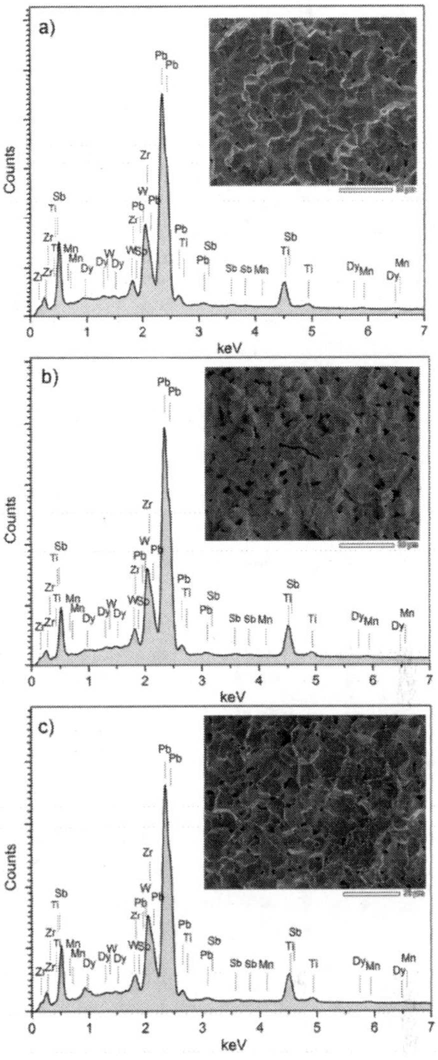

Figure 3. The EDS analysis image of the element distribution for the multicomponent PZT-type ceramics: a) PZT-1, b) PZT-2, c) PZT-3.

3.3 Dielectric properties and DC conductivity

Temperature measurements of the dielectric permittivity of the multicomponent PZT-type materials at frequency 1 kHz are presented in Fig. 4a. All obtained compositions show high values of dielectric permittivity with a clear, sharp phase transition (from the ferroelectric phase to the paraelectric one). At room temperature, the values of the dielectric permittivity are in the range of 447 to 848 (Table 1).

The highest values of the dielectric permittivity maximum show the PZT-2 ceramic sample (with intermediate amounts of tungsten and dysprosium admixtures in its chemical composition). From the group of obtained compositions the lowest values of the dielectric permittivity maximum are shown by the composition of PZT-1 (with the largest amount of dysprosium and the smallest amount of tungsten).

The multicomponent PZT-type ceramics are characterized by low dielectric loss (Fig. 4b). In the area of phase transition, there is a local maximum of the dielectric loss. It is characteristic for multicomponent PZT materials with a perovskite-like structure. Above temperature of 350°C there is a rapidly increase in dielectric loss associated with an increase in electrical conductivity at higher temperatures.

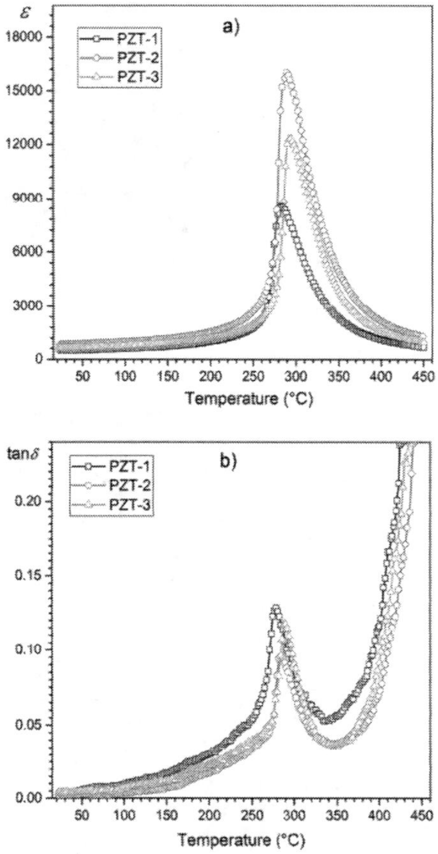

Figure 4. Temperature dependencies of the $\varepsilon(T)$ (a) and $\tan\delta(T)$ (b) for multicomponent PZT-type ceramics (a heating cycle), at frequency 1 kHz.

The frequency dependencies (in the frequency range from 20 Hz to 1 MHz) of dielectric permittivity at various temperatures (from 25°C to 400°C) of the multicomponent PZT-type ceramics are shown in Fig. 5. The values of dielectric permittivity decreases with the increase of frequency (in frequency region from 20 Hz to 5 kHz), and it attains a constant value at high frequencies. This phenomenon is typical for a Maxwell-Wagner type of interfacial polarization.

Fig. 6 shows the temperature dependencies of $\ln\sigma_{DC}(1000/T)$ for the multicomponent PZT-type materials. The DC electrical conductivity tests showed that all PZT-type ceramics have similar waveforms. As the temperature increases, the direct current conductivity increases, too. The research showed no clear trend related to the variable amount of tungsten and dysprosium in the basic PZT-type composition.

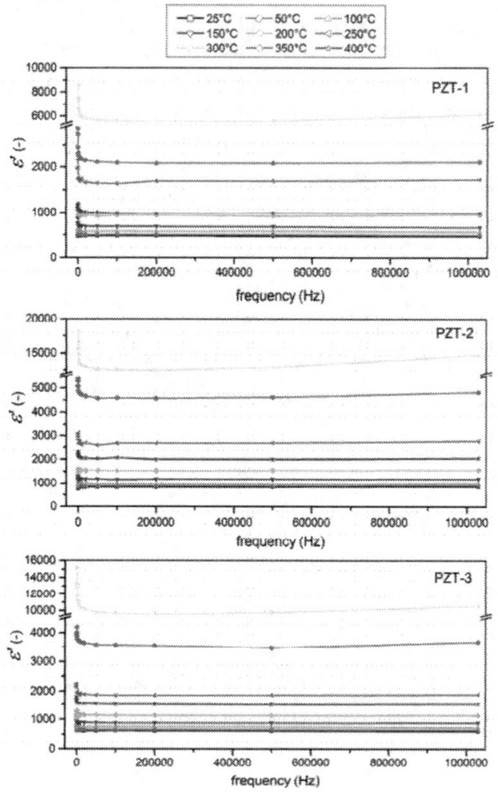

Figure 5. Frequency dependencies of dielectric permittivity of the multicomponent PZT-type ceramics.

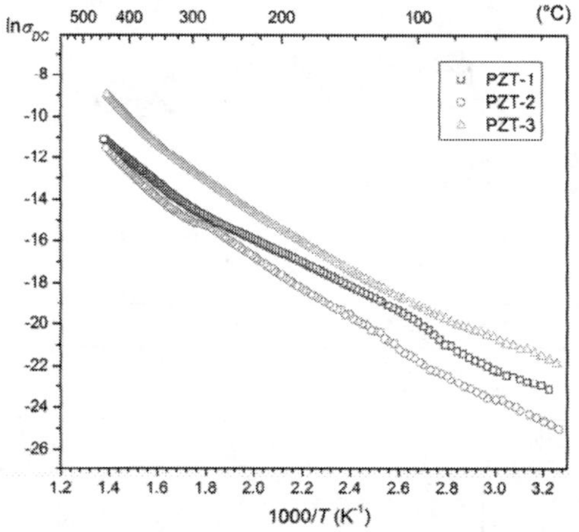

Figure 6. The $\ln\sigma_{DC}(1000/T)$ dependencies multicomponent PZT-type ceramics.

4 Conclusion

In paper three multicomponent PZT-type ceramics doped by Mn^{4+}, Sb^{3+}, Dy^{3+}, W^{6+} have been designed and obtained at following technological conditions: T_s= 1170°C/t_s=2 h.

Results obtained from the X-ray diffraction confirmed single-phase state of the PZT with the perovskite type structure and tetragonal crystallographic system. The microstructures of the multicomponent PZT-type compositions prove that the technological process has been properly carried out and correctly selected technological conditions (ceramic grains exhibit a correct angular shape). All obtained compositions show high values of dielectric permittivity with a sharp ferroelectic-paraelectric phase transition as well as very low dielectric loss.

The good electrophysical properties of the multicomponent PZT-type ceramics, allowed using this type materials as an element to build actuators and piezoelectric transducers in modern micromechatronic and microelectronic devices.

References

1. Z.-G. Ye, *Handbook of Advanced Dielectric, Piezoelectric and Ferroelectric Materials. Synthesis, Properties and Applications* (Woodhead Publishing and Maney Publishing, Cambridge 2008)
2. P.-H. Xiang, X.-L. Dong, H. Chen, Z. Zhang, J.-K. Guo, Ceram. Int. **29**, 499 (2003)
3. V.R. Mudinepalii, R.K. Sirugudu, N. Ramamanohar Reddy, K.V. Sivakumar, V.R.K. Murthy, B.S. Murty, Mater. Chem. Phys. **126**, 295 (2011)
4. J. Li, Q. Sun, Rare Metals **27**, 362 (2008)
5. B. Jaffe, R. Cook, H. Jaffe, *Piezoelectric Ceramics*, (Academic Press, New York, 1971)
6. H. Wei, Y. Chen, Ceram. Int. **41**, 6158 (2015)
7. W. Qiu, H. H. Hng, Mater. Chem. Phys. **75**, 151 (2002)
8. D. Bochenek, P. Niemiec, M. Adamczyk, R. Skulski, R. Zachariasz, B. Wodecka-Duś, Z. Machnik, Arch. Metall. Mater. **62**, 667 (2017)
9. E. Boucher, B. Guiffard, L. Lebrun, D. Guyomar, Ceram. Int. **32**, 479 (2006)
10. L. Kozielski, M. Adamczyk, N. Schreithofer, W. Sakamoto, R. Nowak, Jpn. J. Appl. Phys. **47**, 2176 (2008)
11. R. Zachariasz, D. Bochenek, K. Dziadosz, J. Dudek, J. Ilczuk, Arch. Metall. Mater. **56**, 1217 (2011)
12. D. Bochenek, P. Niemiec, R. Zachariasz, A. Chrobak, G. Ziółkowski, Arch. Metall. Mater. **58**, 1013 (2013)
13. A. Bouzid, E.M. Bourim, M. Gabbay, G. Fantozzi, J. Eur. Ceram. Soc. **25**, 3213 (2005)
14. N. Izyumskaya, Y. I. Alivov, S. J. Cho, H. Morkoç, H. Lee, Y. S. Kang, Crit. Rev. Solid State **32**, 111 (2007)
15. A.J. Moulson, J.M. Herbert, *Electroceramics: Materials, Properties, Applications* (second ed., Wiley, Chichester, West Sussex, 2003)
16. N. Texier, Ch. Courtois, M. Traianidis, A. Leriche, J. Eur. Ceram. Soc. **21**, 1499 (2001)
17. M. Morawiec, A. Grajcar, Journal of Achievements in Materials and Manufacturing Engineering **80**, 11 (2017)

Dielectric properties of the PFN ceramics obtained different chemical-wet technology and sintering by hot pressing method

Dariusz Bochenek[1] and Przemysław Niemiec[1,a]

[1] *University of Silesia in Katowice, Faculty of Computer Science and Material Science, Institute of Technology and Mechatronics 12, Żytnia St., 41–200, Sosnowiec, Poland*

Abstract. The paper shows three kinds of chemical-wet technology synthesis of the ferroelectric-ferrimagnetic $PbFe_{1/2}Nb_{1/2}O_3$ materials and technology process of the ceramic samples sintered by hot pressing method. In the technological process of the PFN powders were used follows precursors: lead acetate trihydrate and niobium ethoxide while an iron-related components were changed: iron nitrate obtained by first sol-gel technology (PFN-n), iron oxalate obtained by second sol-gel technology (PFN-ox) and iron citrate obtained by third technology (PFN-c). XRD, SEM tests and temperature measurements of dielectric properties of the ceramic samples were carried out. The SEM microstructure tests of the PFN ceramic samples confirm the properly conducted technological processes. The best dielectric properties of the PFN ceramic samples were obtained for third technology using iron citrate precursor. But, this method of the PFN ceramic powder requires minimization of a pyrochlore phase and reduction porosity of the ceramic samples.

1 Introduction

Multiferroics are smart materials which show simultaneously at least two types of physical (ferroic) states e.g. a ferromagnetic state (antiferromagnetic, ferrimagnetic), a ferroelectric state (antiferroelectric, ferrielectric), a ferroelastic state (ferromagneticelastic, ferroelastoelectric), ferrotoroid state and showing coupling between subsystems [1-4]. Ceramic materials with multiferroic properties found functional application as various types of piezoelectric transducers, electric bandfilters, sensors, generators, servomotors, phase modulators, frequency multipliers, electromechanical, electroacoustic, pyroelectric transducers, memory elements, etc. [5-7]. These tape multiferroic materials are designed in the form of solid solutions [8] and multicomponent compounds [9-10], thin films [11], and composites [12-15].

The ferroelectromagnetic $PbFe_{1/2}Nb_{1/2}O_3$ ceramics, shortly called PFN, is one of the materials belonging to group of multiferroics. The PFN ceramics has perovskite structure with the general $A(B_{1-x}B_x)O_3$ formula, where Pb^{2+} in site A and Nb^{5+} in site B favour electrical order and Fe^{3+} in the B site gives rise to a magnetic moment that generates magnetic order [16].

Many methods and techniques for obtaining PFN ceramic powders are known e.g. a molten salt synthesis, a reaction in solid state, two-stage columbite, sol–gel, co-precipitation method, wolframite route, high-energy milling and others [17-20]. Each of these methods requires a used of specific technological conditions and diligence in laboratory process in order to obtained materials with good properties.

The aim of the work was obtained and studied the ferroelectromagnetic PFN powders by liquid phase synthesis and sintered ceramic samples by hot pressing method. The work show the influence of the synthesizing sol-gel method used on the electrophysical properties of PFN ceramic samples.

In the paper the structural, microstructural and dielectric properties of the PFN samples were investigated.

2 Experiment

2.1 The first technology of the PFN material

In the first chemical-wet sol-gel technology the $PbFe_{1-x}Nb_xO_3$ powder was obtained using as precursors lead (II) acetate trihydrate $Pb(CH_3COO)_2{\cdot}3H_2O$ (POCH), iron (III) nitrate nonahydrate $Fe(NO_3)_3{\cdot}3H_2O$ (POCH), niobium (V) ethoxide $Nb(OC_2H_5)_5$ (Aldrich). In the process 2-metoxyethylate was a solvent and acetylacetone was a stabilizer (used both to control a hydrolysis of alcoholates and to protect against too quick gelation of the solution) while distilled water was used for the hydrolysis reaction. The lead acetate was solubilized in the 2-methoxyethylate by mixing and heating the solution. The iron nitride and the niobium ethylate were solubilized in 2-methoxyethylate, too. All components were weighed in a stoichiometric way (the

[a] przemyslaw.niemiec@us.edu.pl

lead (II) acetate trihydrate was added in excess because of its low evaporation temperature). The next step of the sol-gel technology was to mix obtained earlier solutions with their simultaneous heating. Afterwards the acetylacetone was added to the solution as a stabilizer as well as distilled water was added to start a hydrolysis process (a sol is formed). In order to hasten gelation process and to change the gel into a powder the solution was annealed under the IR lamp. In detail, the technological process was presented at work [21].

The PFN powder obtained after drying was mixed and to remove organic parts the calcined process was carried out. The process of remove organic parts (T=600°C for t=2 h) proceeds in a turbulent manner (gases are released from a powder in the form of brown smoke). The gelation process of the PFN material is shown in Fig. 1. The PFN ceramic samples obtained in first technology from iron (III) nitrate as PFN-n were marked.

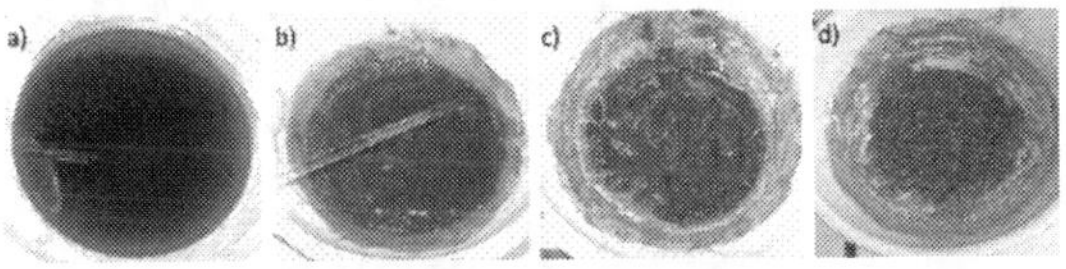

Figure 1. The gelation process of the PFN material in first sol-gel technology (PFN-n); a) solution of the PFN, b) the gelation process, c) the gelation process supported by IR rays, d) PFN-n powder.

2.2 The second technology of the PFN material

In the second sol-gel technology the $PbFe_{1-x}Nb_xO_3$ powder was obtained using as precursors lead (II) acetate $Pb(CH_3COO)_2{\cdot}3H_2O$ (POCH), niobium (V) ethoxide $Nb(OC_2H_5)_5$ (Aldrich) and iron (III) oxalate $FeC_2O_4{\cdot}3H_2O$ (Aldrich). The lead (II) acetate $Pb(CH_3COO)_2{\cdot}3H_2O$ was dissolved in the ethyl glycol heating and mixing. The niobium ethylate $Nb(OC_2H_5)_5$ was prepared in a similar way. In the case of the iron (III) oxalate a small amount of the ethyl glycol (POCH) and a small amount of a nitric acid was added and mixture was heated intensively. It results in a transition of the iron in the compound from II to III oxidation degree. In the next step, the three obtained component were mixed together and heating. After this, the acetylacetone was added (continuously heating) and the distilled water to start a hydrolysis process. In order to accelerate sol and gelation processes IR rays were used. In detail, the technological process was presented at work [22]. The gelation process of the PFN material is shown in Fig. 2.

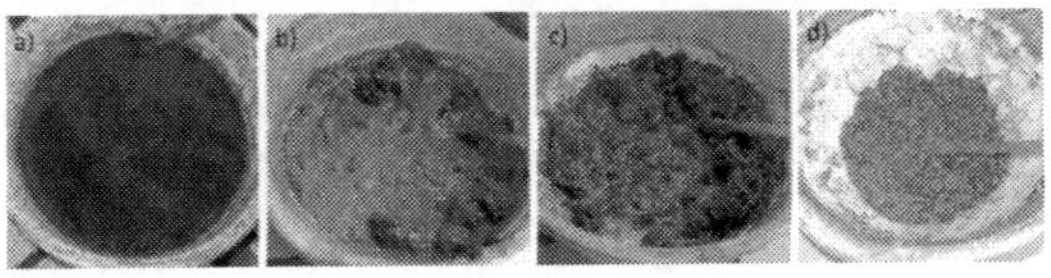

Figure 2. The gelation process of the PFN material in second sol-gel technology (PFN-ox); a) solution of the PFN, b-c) the gelation process supported by IR rays, d) PFN-ox powder.

After drying the PFN powder was mixed and next removes organic parts was done by calcinated method at the following conditions 600°C/2 h. The PFN ceramic samples obtained in second technology from iron (III) oxalate as PFN-ox were marked.

2.3 The third technology of the PFN material

In the third chemical-wet technology (solution precipitation method) the $PbFe_{1-x}Nb_xO_3$ powder was obtained using as precursors ferric citrate monohydrate $C_6H_5FeO_7{\cdot}H_2O$ (Fluka), lead (II) acetate trihydrate $Pb(CH_3COO)_2{\cdot}3H_2O$ (POCH) and niobium (V) ethoxide $Nb(OC_2H_5)_5$ (Aldrich). The solution precipitation method used in PFN technology is a variation of the sol-gel method. The $Pb(CH_3COO)_2{\cdot}3H_2O$ was dissolved in the CH_3COOH acetic acid and simultaneously to $Nb(OC_2H_5)_5$ ethylene glycol (POCH) with 2-methoxy-ethanol (Fluka) was added. Subsequently, to the ferric citrate $C_6H_5FeO_7{\cdot}H_2O$ the distilled water was added and then the solution was heated. After this, the acetic acid was added to increase effectiveness of dissolving and it was brought to boiling (still stirring intensively). In the next step, all the component solutions were mixed together and heated simultaneously (this initiates the precipitation process), and the acetylacetone was add. Subsequently, the distilled water was added for the hydrolysis process. The obtained liquid solution was dried under an IR lamp in order to vaporize the solvent and to obtain ceramic powder.

Similar to the previous methods, after drying the PFN powder was mixed and removes organic parts was done by calcination route at following conditions: 600°C/3 h. In detail, the technological process was presented at work [23]. The process obtaining of the PFN material is shown in Fig. 3 and the PFN ceramic samples obtained in third technology from ferric citrate as PFN-c were marked.

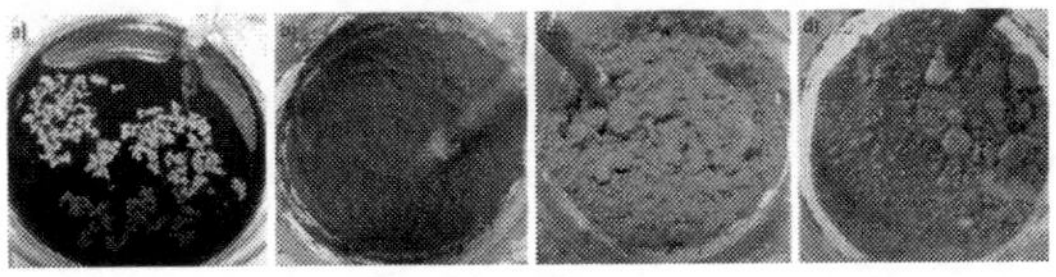

Figure 3. The co-precipitation method of the PFN material (PFN-c); a) the solution with precipitation process, b-c) the solvent vaporization supported by IR rays, d) PFN-c powder.

2.4 Sintering of the PFN powders and samples preparation

All the PFN powders obtained with three technologies after mixing process were pressed into the ceramic pellets. Densification was made by hot pressing method (HP) at following conditions: sintering temperature T_s=1000°C, holding time t_s=1 h and the pressure p_s=20 MPa.

The synthesized PFN powders were pressed into pellets with a thickness of 1.0 mm. The mechanical stress was removed by annealing process at following condition 750°C/0.25 h, while for the electrical measurements the surfaces of the pellets were covered with the silver electrode (850°C/0.25 h).

https://doi.org/10.1051/matecconf/201824201002

2.5 Characterization

The X-ray powder-diffraction data (XRD) of the sintered PFN samples were collected at room temperature (*RT*) on a diffractometer Phillips X'Pert APD (Cu-Kα radiation). The data were collected in the 2θ range from 10° to 65° in steps of 0.02 degrees with an integration time of 4 s/step. Microstructure of the ceramic samples was observed by scanning electron microscopy JSM-7100F TTL LV. The stoichiometry of the PFN ceramics was studied with the EDS chemical composition analysis system. Prior to the SEM/EDS analyses the samples were coated with gold to provide electrical conductivity and to avoid any charging effects.

Dielectric measurements were carried out on a capacity bridge of a QuadTech 1920 Precision LCR Meter for a heating cycle (in temperature range from 20°C to 250°C) and the measurements field frequency in the range from 100 Hz to 100 kHz.

3 Results and discussion

3.1. X-ray diffraction analysis

The X-ray powder-diffraction patterns at room temperature of PFN materials are presented in Fig. 4. The XRD measurements show that obtained materials have a perovskite structure and all samples are matching well with the tetragonal symmetry with *P4mm* space group (JCPDS card no. 04-009-5124). The XRD tests show that in the case of the PFN-c sample beside perovskite phase material has too, small amount of the pyrochlore phase ($Pb_2Nb_2O_7$). In case of the PFN-n sample there are very small amount foreign phase, while in the case of the PFN-ox there is no foreign (pyrochlore) phase.

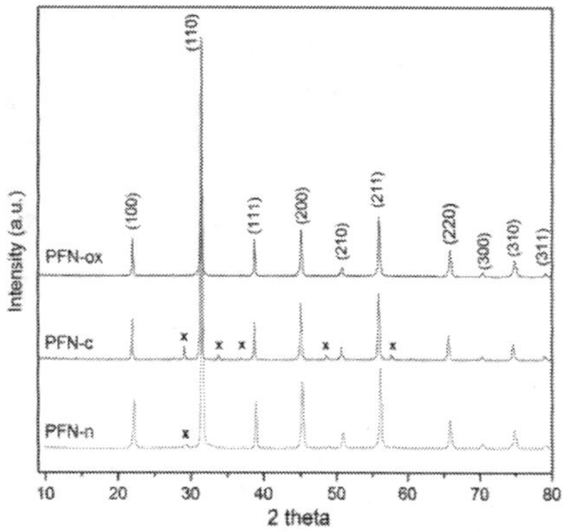

Figure 4. X-ray diffraction patterns of the PFN materials.

The percentage of perovskite phase in the PFN materials was calculated from the following equation (1):

$$P_{phase} = [I_{perov}/(I_{perov}+I_{pyroch})]\cdot 100\% \quad (1)$$

where: I_{perov} is the intensity of highest perovskite peak (110) and I_{pyroch} is the intensity of highest pyrochlore peak (222). Percentage of perovskite phase contents in case of the PFN-c sample is 94.4 %, whereas in case of the PFN-n sample is 97.86 %.

3.2. Microstructural and EDS tests

The images of morphology of fracture surfaces of the PFN samples obtained with different technologies are presented in Fig. 5. The microstructure of the PFN-n sample shows grain size heterogeneity. In the entire sample an angular of the ceramic grains is not fully correct.

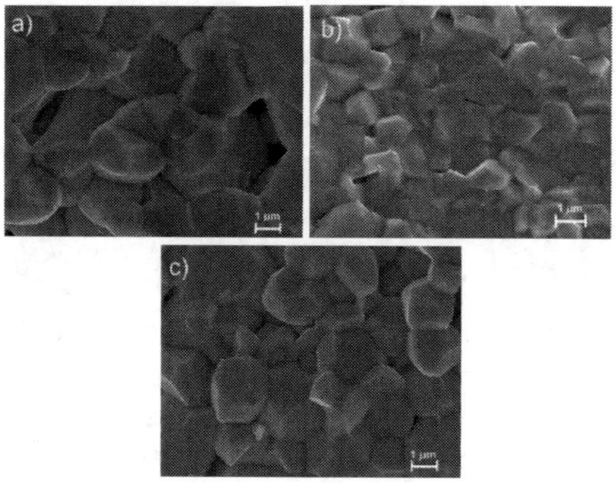

Figure 5. SEM images of the PFN ceramics: a) PFN-n, b) PFN-ox, c) PFN-c.

The breakthrough of the PFN-n sample is mixed and predominantly occurs through grain boundaries. However, when the fracture occurs through the grain, the surface morphology is solid, without any small elements originating from the cracked grain. This behavior indicates high mechanical strength of the inside of the grains and confirms the properly conducted technological process. The PFN-ox sample is characterized by a fine-grained microstructure and densely packed, the grains of which have regular edges. The breakthrough of the PFN-ox sample occurs predominantly through the grain, which breaks in a compact solid manner (grain with a strongly consolidated structure does not break down into small elements). The PFN-ox ceramics have a fine closed porosity.

Among the ceramic samples obtained, the PFN-c one has the best microstructure. The grains are properly crystallized, and the microstructure shows high grain homogeneity. As in previous cases, there are closed pores in the volume of sample. The numerous grain agglomerates formed during the technology promote the formation of closed porosity. The sample breakthrough occurs mainly through grain boundaries, which is confirmed by properly selected technological sintering conditions of the ceramic ensuring high mechanical strength of the grain interior.

The EDS analysis (Fig. 6) confirmed the presence of constituent elements in the test PFN ceramic samples without the presence of foreign elements. In the Table 1 the percentage of the individual components of the PFN ceramics were given. Theoretical calculations were made on the basis of the chemical formulas of the obtained materials. In the case of the PFN-n sample, a small amount of iron and niobium deficiency and excess of lead are observed compared to theoretical calculations. In the case of the PFN-ox sample, a small amount of niobium deficiency and excess of iron has been demonstrated. In the case of the PFN-c sample lead and niobium deficiency and excess of iron are observed. For this technology percentage deviation from the theoretical composition are the largest.

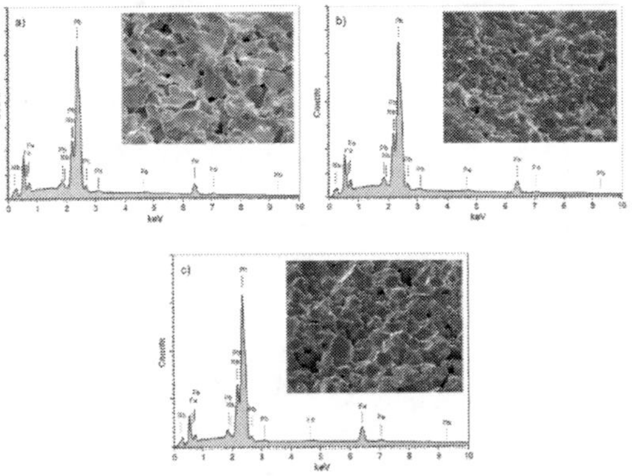

Figure 6. EDS testes of the PFN samples: a) PFN-n, b) PFN-ox, c) PFN-c.

Table 1. Theoretical and experimental percentages of elements of the PFN samples.

	PFN	PFN-n	PFN-ox	PFN-c
	Theor. (%)	Exptl. (%)	Exptl. (%)	Exptl. (%)
PbO	67.72	69.34	67.84	65.34
Fe_2O_3	12.11	11.98	12.89	15.96
Nb_2O_5	20.16	18.68	19.27	18.71

3.3 Dielectric properties

For the PFN ceramics, the maxima of dielectric permittivity (Fig. 7) are not as "sharp" as in the case of the ferroelectric single crystals or multicomponent PZT-type ceramics in which the phase transition occurs at a strictly defined temperature (Curie temperature) [24-25].

At room temperature (*RT*), the PFN-n ceramics have high values of dielectric permittivity (for 1 kHz ε=2,700). On the temperature dependencies $\varepsilon(T)$ the graphs for the lower frequencies, do not show clear maximum dielectric permittivity occurring at the phase transition temperature. At higher frequencies (over 2 kHz), the characteristic maximum becomes clearer with broad form.

Of the samples tested, the PFN-ox ceramics have the highest values of dielectric permittivity and show a broad phase transition occurring over a wider range of temperatures. On the temperature dependencies $\varepsilon(T)$ with increasing frequency, frequency dispersion and a sudden drop in the dielectric permittivity are observed. For 1 kHz, the permittivity value maximum (ε_{max}) is 26,840 (at T_m) and approx. 2,600 (at *RT*). In comparison with the PFN ceramics obtained by free sintering method (at following conditions: T_s=1100°C and T_s=1150°C) presented in the work [22], the PFN-ox ceramics have slightly lower ε_{max} values.

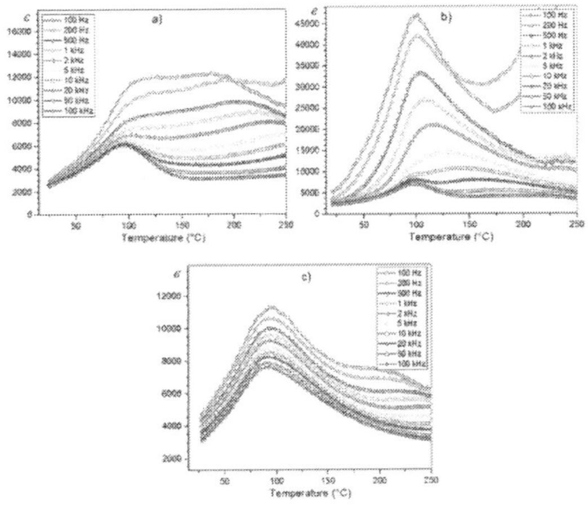

Figure 7. Temperature dependencies of dielectric permittivity (ε) for the PFN ceramics: a) PFN-n, b) PFN-ox, c) PFN-c.

The PFN-c ceramics have a broad phase transition but occurring within a narrow temperature range. There is no frequency dispersion with the frequency increase (the characteristic maximum phase transition occurs at the same temperature). From the tested PFN samples, the values of dielectric permittivity at *RT* are the highest. For 1 kHz, the value of the dielectric permittivity maximum (at T_C) is 9,590 and at *RT* ε=4,100. Compared with the PFN ceramics obtained by FS [23], the PFN-c ceramics have a lesser degree of blurring the phase transition and higher values of ε_{max}. In the work [23] the value of ε_{max} is 5,110.

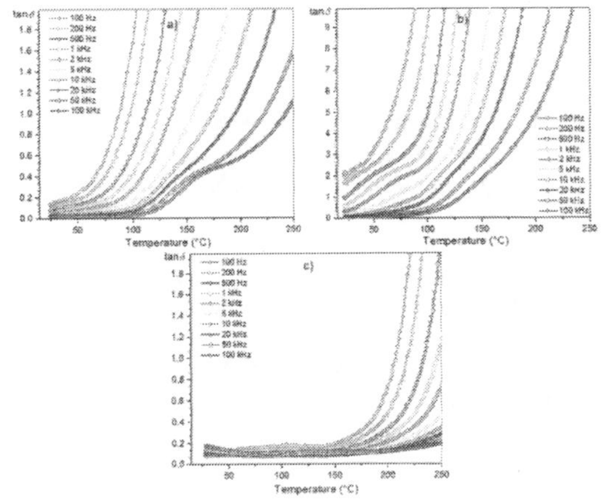

Figure 8. Temperature dependencies of dielectric loss tangent (tanδ) for the PFN ceramics: a) PFN-n, b) PFN-ox, c) PFN-c.

Fig. 8 presents the results of temperature measurements of dielectric loss for PFN ceramic samples.

In the case of the PFN-n ceramics with increasing temperature, dielectric loss increase. Just before the phase transition temperature, the values of the tanδ increase rapidly. At *RT* (1 kHz), value of the dielectric loss is 0.040. In comparison with the PFN ceramics obtained by HP at higher temperatures (T_s=1050°C and T_s=1100°C) presented in the work [21], the PFN-n ceramics have lower value of dielectric loss.

Comparing the obtained PFN samples, the PFN-ox ceramics have the highest dielectric loss. The high values of the tanδ occur both at low temperatures (0.58 for 1 kHz) as well as at higher temperatures, where the increase in dielectric loss is very high. In comparison with the PFN ceramics obtained by FS method [22], the PFN-ox ceramics show slightly higher dielectric loss.

The PFN-c ceramics exhibits the lowest dielectric loss over the entire temperature and frequency measurement area. At *RT* (1 kHz), dielectric loss is 0.086, while at T_m, 0.099. Low dielectric loss may be associated with the presence of more pyrochlore phase (as shown by XRD tests), which partially reduces the electrical conductivity from the intergranular areas, reducing values of dielectric loss. In comparison with the PFN ceramics obtained by FS method [23], the PFN-n ceramics have significantly lower dielectric loss.

4 Conclusion

The paper shows the technological process of the three kinds of chemical-wet method synthesis of the multiferroic $PbFe_{1/2}Nb_{1/2}O_3$ materials and sintered by hot pressing method. Differences between individual technologies were related to the change in Fe component, i.e. iron nitrate obtained by first sol-gel technology (PFN-n), iron oxalate obtained by second sol-gel technology (PFN-ox) and iron citrate obtained by third technology (PFN-c).

The XRD measurements show that despite the pyrochlore phase, the PFN-c ceramics have the best dielectric properties, i.e. the optimum of the dielectric permittivity and low dielectric loss, compare to the all obtained PFN ceramic samples. SEM microstructure tests of the PFN ceramic samples confirm the properly conducted technological processes.

Comparing three technologies of the PFN syntheses performed by the sol-gel method, it can be concluded that the optimal technology is the process using iron citrate (third method). However, this process requires elimination or minimization of a pyrochlore phase in the PFN ceramics as well as reduction of the porosity ceramic samples (eliminating creation of hard clusters agglomerate of the powder that promote the formation of significant closed porosity). Minimizing the possibility of powder agglomeration formation can be partially eliminated by breaking them with ultrasound methods. The results of such research will be the subject of the next work of the authors.

Acknowledgements

The authors thank the Dr Katarzyna Osińska for help in the technological process.

References

1. D. Khomskii, Physics **2**, 20 (2009).
2. H. Schmid, J. Phys.: Condens. Matter **20**, 434201 (2008)
3. K.F. Wang, J.-M. Liu, Z.F. Ren, Adv. Phys. **58**, 321 (2009)
4. M.H. Lente, J.D.S. Guerra, G.K.S. de Souza, B.M. Fraygola, C.F.V. Raigoza, D. Garcia, J.A. Eiras, Phys. Rev. B **78**, 054109 (2008)
5. D. Bochenek, Z. Surowiak, J. Krok-Kowalski, J. Poltierova-Vejpravova, J. Electroceram. **25**, 122 (2010)
6. J.F. Scott, J. Mater. Chem. **22**, 4567 (2012)
7. Y. Xu, *Ferroelectric Materials and Their Applications*. (Elsevier, North Holland, New York, 1991)
8. D. Bochenek, J. Dudek, Eur. Phys. J.–Spec. Top. **154**, 19 (2008)
9. D. Bochenek, P. Niemiec, P. Guzdek, M. Wzorek, Mater. Chem. Phys. **195**, 199 (2017)
10. X. Meng, K.Z. Baba-Kishi, G.K.H. Pang, H.L. Chan, CH.-L. Choy, H. Luo, Ferroelectrics **297**, 57 (2003).
11. K. Tahmasebi, A. Barzegar, Transaction F: Nanotechnology **17**, 108 (2010)
12. D. Bochenek, P. Niemiec, R. Skulski, A. Chrobak, P. Wawrzała, Mater. Chem. Phys. **157**, 116 (2015).
13. S. Srinivas, J.Y. Li, Acta Mater. **53**, 4135–4142 (2005)
14. J. Kulawik, D. Szwagierczak, P. Guzdek, J. Magn. Magn. Mater. **324**, 3052 (2012)
15. L.M. Hrib, O.F. Caltun, J. Alloys Compds. **509**, 6644 (2011)
16. O. Raymond, R. Font, N. Suárez-Almodovar, J. Portelles, J. M. Siqueiros, J. Appl. Phys. **97**, 084107 (2005)
17. Y.-Ch. Liou, Ch.-Yu Shih, Ch.-H. Yu, Jpn. J. Appl. Phys. **41**, 3829 (2002)
18. S. Ananta, N.W. Thomas, J. Eur. Ceram. Soc. **19**, 2917 (1999)
19. S.B. Majumder, S. Bhattacharyya, R.S. Katiyar, A. Manivannan, P. Dutta, M.S. Seehra, J. Appl. Phys. **99**, 024108 (2006)
20. X. Gao, J. Xue, J. Wang, T. Yu, Z.X. Shen, J. Am. Ceram. Soc. **85**, 565 (2002).
21. D. Bochenek, Z. Surowiak, J. Alloy. Compd. **480**, 732 (2009)
22. D. Bochenek, R. Zachariasz, Materialwiss. Werkst. **46**, 294 (2015)
23. D. Bochenek, J. Alloy. Compd. **504**, 508 (2010)
24. L.B. Kong, J. Ma, H.T. Huang, W. Zhu, O.K. Tan, Mater. Lett. **50**, 129 (2001)
25. B. Wodecka-Duś, D. Czekaj, Arch. Metall. Mater. **54** 923 (2009)

Synthesis and structure of $BiFeO_3$:RE (RE=Gd^{3+}, Dy^{3+}, Nd^{3+}) ceramics

J. Dzik[1,a], M. Płońska[1], T. Pikula[2], M. Rerak[1]

[1] *University of Silesia in Katowice, Faculty of Computer Science and Material Science, Institute of Technology and Mechatronics 12, Żytnia St., 41–200, Sosnowiec, Poland*

2 Lublin University of Technology, Institute of Electronics and Information Technology, 38A Nadbystrzycka Str., 20-618 Lublin, Poland

Abstract. In the present work the influence of rare earth elements concentration (0-10at-%) on $BiFeO_3$:RE (RE=Gd^{3+}, Dy^{3+}, Nd^{3+}) ceramics were studied. All ceramic powders were synthesized by conventional ceramic method using high purity raw materials (>99,9%), and subsequently sintered by free sintering and cold pressing method. To analyze the powders and ceramics more the XRD, EDS, SEM, and DTA were performed.

1 Introduction

Multiferroic materials are the subject of intensive research during recent years, due to their interesting physical properties [1], as well as the great technological potential of the materials for microelectronics and spintronics [2,3]. In these materials in which ferro/antiferromagnetic and ferroelectric properties occur simultaneously. Bismuth ferrite and its materials are attracting great attention owing to both great technological potential and the interesting physics behind their functional properties [4]. $BiFeO_3$ ceramic is an interesting candidate due to its high ferroelectric curie temperature (T_C=850°C) and antiferromagnetism below Neel temperature (T_N=370°C) [5,6]. These characteristics are unrepeatable, taking into account that most magnetic ferroelectrics possess coexisting spin and dipole order only well below room temperature. At room temperature, $BiFeO_3$ single crystal has distorted rhombohedral (R) structure with lattice parameter of (a_r=3.96Å and a_r=89.41 Å) [7]. The magnetic ordering is G-type with a weak canting moment with a 62 nm spin cycloid [8]. When discussing the properties of bismuth irons, it should be noted that information on the $BiFeO_3$ formations and temperature stability limits remains ambiguous. Despite the fact that bismuth ferrite is characterized by very good physical properties, attempts are made to improve the ferroelectric and ferromagnetic properties. One possible strategy for obtaining improved properties in BiFeO3 is partial ionic substitution. Additionally, the pure, single-phase bismuth ferrite is typically very difficult to obtain via solid-state reaction of Bi2O_3 and Fe_2O_3 [9]. During its preparation additional (impurity) phases such as $Bi_2Fe_4O_9$, $Bi_{36}Fe_{24}O_{57}$ or $Bi_{25}FeO_{39}$ appear very often [10,11]. Lomanova et al [12] reported the possibility of $Bi_2Fe_4O_9$ and $Bi_{25}FeO_{39}$ formation during $BiFeO_3$ synthesis and it was shown to be dependent upon the quality of the initial reagents [19]. They argue that not enough pure precursors can result in the formation of the above-mentioned phases and to their stable existence as impurities during $BiFeO_3$ formation. Other authors postulate that it is difficult to synthesize the single-phase $BiFeO_3$ because $Bi_2Fe_4O_9$ and $Bi_{25}FeO_{39}$ are thermodynamically more stable than $BiFeO_3$. It has been observed that the partial substitution of Bi atoms by elements like Eu, La, Nd, Sm, Tb [13,14,15] allows to eliminate the impurity phases. The short review of the results of experimental investigations of bismuth ferrite shows the necessity of performing more comprehensive and deeper research. Special attention should be paid to rare-earth substituted compounds, whose properties remain poorly known what can be stated after reading the available literature. Aim of this study was to synthesize and fabricate $Bi_{1-x}A_xFeO_3$ ((A = Nd, Gd, Dy, for x = 0.03, 0.05, 0.07, 0.1) ceramics by solid state reaction. By means of simultaneous thermal analysis (TG / DTG, DTA) and X-ray diffraction analysis the process of synthesis of $Bi_{1-x}A_xFeO_3$ ceramics has been studied.

2 Experimental

2.1 The technology of the PFN material

The preparation and synthesis process of undoped and Gd^{3+}, Dy^{3+} and Nd^{3+} co-doped $BiFeO_3$ powders was performed through the solid state reaction, by the conventional ceramic method. Stoichiometric amounts of high purity oxides powders, Bi_2O_3 (Sigma – Aldrich,

[a] jolanta.dzik@us.edu.pl

99,9%), Fe_2O_3 (Sigma – Aldrich, 99,9%), Nd_2O_3 (Sigma – Aldrich, 99,9%), Gd_2O_3 (Sigma – Aldrich, 99,9%), Dy_2O_3 (Sigma – Aldrich, 99,9%) were weighed according to the nominal composition of $Bi_{1-x}A_xFeO_3$. The appropriate quantities of reagents were weighed according to the formula (1):

$$\left(\frac{1-x}{2}\right) Bi_2O_3 + \left(\frac{x}{2}\right)A_2O_3 + \left(\frac{1}{2}\right)Fe_2O_3 \rightarrow Bi_{1-x}A_xFeO_3 \quad (1)$$

The high purity (99.9%) oxide powders were ground homogeneously in the mortar at room temperature for 1 h. Next, these mixtures were subjected to grind in a ball mill for 24 h using YTZ balls as grinding media in ethanol solution. The dried mixture of powders was compacted into pellets of 20 mm in diameter by pressing under pressure of p=60 MPa in a stainless – steel die. The synthesis was carried out at T =800^0 C in corundum crucible with air atmosphere for 3h. Calcined material were remilled for 24 h to reduce the particle size, and then cold pressed into pellets (d = 10 mm, p = 30 MPa). Then the material was sintered in air at the temperature of T = 860^{0}C for 3h. The flowchart of the complete fabrication process is shown in Fig. 1.

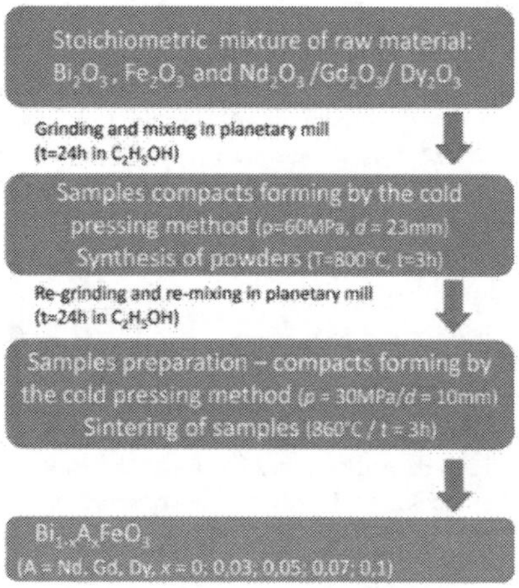

Fig. 1 The flowchart of fabrication process of $Bi_{1-x}A_xFeO_3$ ceramics.

2.2 Characterization

Microstructure and chemical composition of the final ceramics were investigated by scanning electron microscope (SEM) JSM - 7100F equipped with an energy dispersive spectrometer (EDS) NORAN Vantage. The microscope was operating at 15 kV acceleration voltage. Structure of the samples was investigated using PANAlytical X-Pert Pro difractometer with Cu lamp (λ = 0.154 nm). X'Pert HighScore Plus computer program equipped with the ICDD PDF2 data base was used to phase analysis and Rietveld refinement of the structure. The densities of the sintered pellets were measured by Archimedes' principles. Parameters of the thermal treatment were determined by simultaneous thermal analysis (DTA/TG/DTG). Simultaneous measurements were executed by heating the dried powders in air at 10°C/min. The test specimens were prepared in powder form, the reference material was aluminum oxide Al_2O_3. The simultaneous thermal analysis method enables to determine the optimum synthesis temperature.

3 Results and discussion

3.1 Density

Fig. 2 shows the density of obtained ceramic samples $Bi_{1-x}A_xFeO_3$. Density of the samples is closely related to the type of dopant and its content in the material. For the dopant neodymium and dysprosium it can be concluded that with increase of the dopant content in the studied ceramics its density rises too. Vice versa in the case of dopant gadolinium: with increase of the dopant content the density of the ceramic decreases.

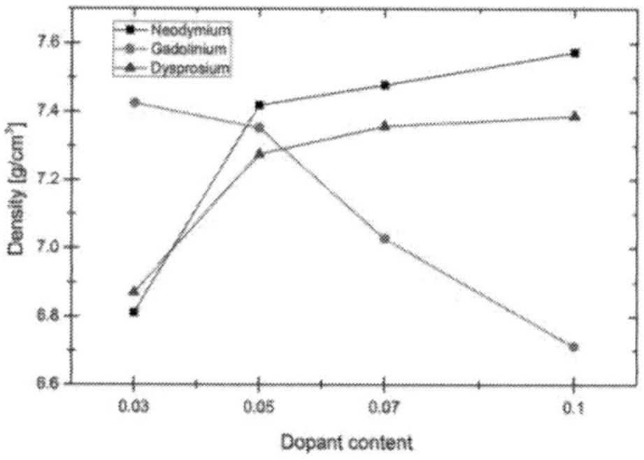

Fig. 2 Effect of dopant content on the density of ceramic samples.

3.2 Simultaneous thermal analysis

Simultaneous thermal analysis, in which thermal analysis and mass change effects are measured concurrently on the same sample was used to investigate synthesis effects in the stoichiometric mixture of powders. A DTA curve obtained for the stoichiometric mixture of Bi_2O_3, Fe_2O_3 and Nd_2O_3 powders for $Bi_{0.9}Nd_{0.1}FeO_3$ is shown in Fig. 3. One can see there are at least two temperature ranges showing rates of the mass loss, namely ΔT_1≈(290-400)°C and ΔT_2≈(580-800)°C. The mentioned mass loss is accompanied by a thermal effect in the form of exothermic peaks in the DTA curve. The DTG curve shows the rate of mass loss as a function of temperature. With the DTG curve it is possible to determine the start and the end of each change-weight temperature region, in other words, the range of the reaction and evaporation of absorbed H_2O and CO_2, as well as slight evaporation of the starting oxides – Bi_2O_3 (in this case), and the reading of the TG curve of the mass change value. Thermogravimetric (TG) plots show that mass loss

https://doi.org/10.1051/matecconf/201824201003

reaches a value of Δm=-1% at T=800°C. The observed weight loss one can ascribe with evaporation of products and/or by-products of chemical reactions as well as evaporation of Bi_2O_3 oxide. At temperature T>800°C no mass change effects occurred. From the practical point of view DTA, TG and DTG curves help to choose the appropriate rate of the thermal treatment of the studied material. The temperature of synthesis was chosen as T=800°C.

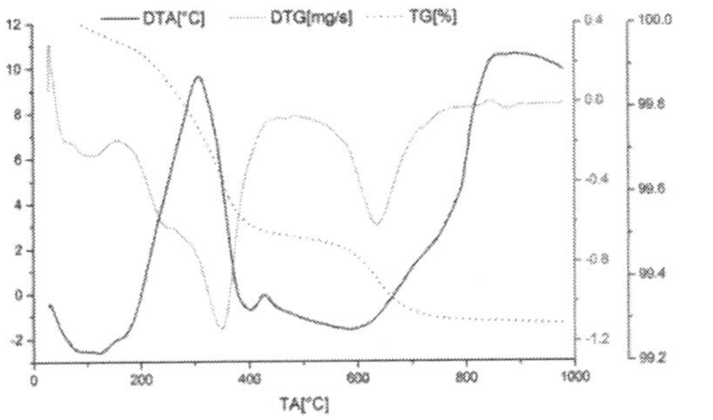

Fig. 3 Differential thermal analysis (DTA) curves for stoichiometric mixtures of oxides used for synthesis of $Bi_{0.9}Nd_{0.1}FeO_3$.

3.1. X-ray diffraction analysis

The results of XRD measurements are presented in Fig. 5. Diffraction lines corresponding to impurity phases ($Bi_2Fe_4O_9$) were marked by symbols whereas all other lines belong to the main phase, i.e. $Bi_{0.9}A_{0.1}FeO_3$ solid solution with a suitable dopant marked on the image. Analysis of the X-ray diffraction patterns of the ceramic powders was carried out using a computer program PowderCell [16]. Refinement of the structural parameters of $Bi_{0.9}A_{0.1}FeO_3$ solid solutions was performed with the Rietveld method e.g. [17,18]. The angular position of diffraction lines agree well with positions for rhombohedral $BiFeO_3$ phase given in ICDD 01-082-1254 card. A small content of $Bi_2Fe_4O_9$ was also noticed (99-100-8872).

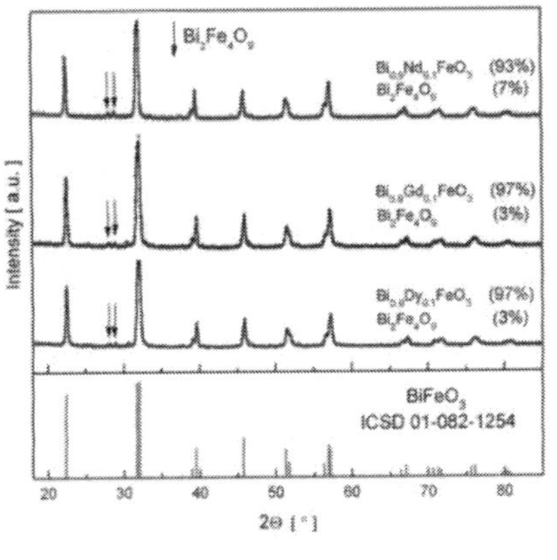

Fig. 4 X-ray diffraction patterns of $Bi_{0.9}A_{0.1}FeO_3$ solid solutions.

On the basis of analysis of diffractograms it was concluded that the structure of the obtained solid materials was rhombohedral (R3c space group). Also the weight fraction of phases from diffractograms was estimated. Diffraction lines coming from impurity phases are rather small. The occurrence of weak lines for 2θ = 27° isn't connected with the $BiFeO_3$-type structure and rather caused by the presence of a vestigial quantity of $Bi_2Fe_4O_9$. For the content of neodymium x=0.1 the concentration of the main phase, i.e. $Bi_{0.9}Nd_{0.1}FeO_3$, is about 93 wt.%. The rest are impurity phases, namely $Bi_2Fe_4O_9$ in amount 7 wt.%. For x = 0.1 of dysprosium the concentration of the main phase, i.e. $Bi_{0.9}Dy_{0.1}FeO_3$, is about 97 wt.%, whereas the impurity phases, namely $Bi_2Fe_4O_9$ in amount 3 wt.%. The weight fraction of phases from diffractograms for the x=0.1 gadolinium are the same like a $Bi_{0.9}Dy_{0.1}FeO_3$ (as shown in Fig.). In another author's work on the influence of rare earth ions on $BiFeO_3$, it is shown that a higher neodymium dopant concentration results in a single phase material [19]. Results of calculations of elementary cell parameters as well as agreement indices are given in Table 1.

Table 1 Structural parameters of $Bi_{0.9}R_{0.1}FeO_3$ (R = Nd, Gd, Dy) derived from Rietveld refinement of XRD patterns.

		$BiFeO_3$+Nd	$BiFeO_3$+Gd	$BiFeO_3$+Dy
Cryst. system		Rhomboh.	Rhomboh.	Rhomboh.
Space group		R3c	R3c	R3c
Lattice param. nm	**a =b**	0.55716(2)	0.55672(3)	0.55645(3)
	c	1.38047(3)	1.38127(5)	1.38174(4)
Volume [nm³]		0.37112	0.37074	0.37052
Agreement indices	R_{exp}	3.9	3.5	3.7
	R_{prof}	4.1	4.8	4.6
	χ^2	1.8	3.0	2.5

The determined lattice parameters from the number of atomic (dopants) are presented in Fig. . The linear drop in the values of a lattice constant may be observed while c parameter increases with the increase of the atomic number.

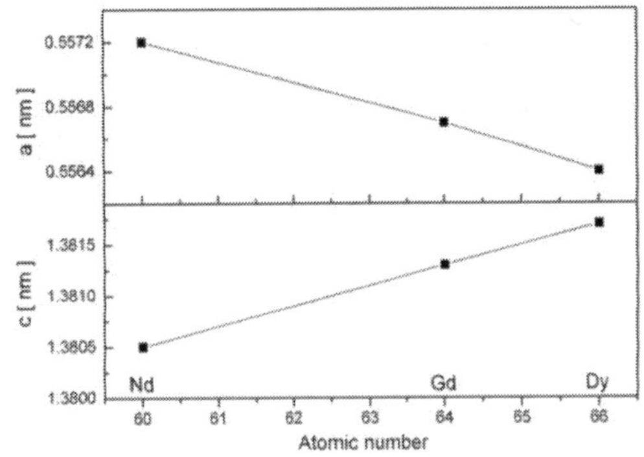

Fig. 5 Lattice constants of $Bi_{0.9}Nd_{0.1}FeO_3$ solid solutions with rhombohedral structure.

3.2. Microstructural and EDS tests

Using scanning electron microscopy analysis microstructure was performed and the chemical compositions of obtained materials were determined. SEM pictures of $Bi_{0.97}A_{0.03}FeO_3$ ceramics obtained by pressureless sintering from stoichiometric mixtures of oxides are shown in Fig.6. Analyzing the SEM image it can be noted, that addition of the individual dopants did not affect the packing of the beans. The shape of the grains changed.

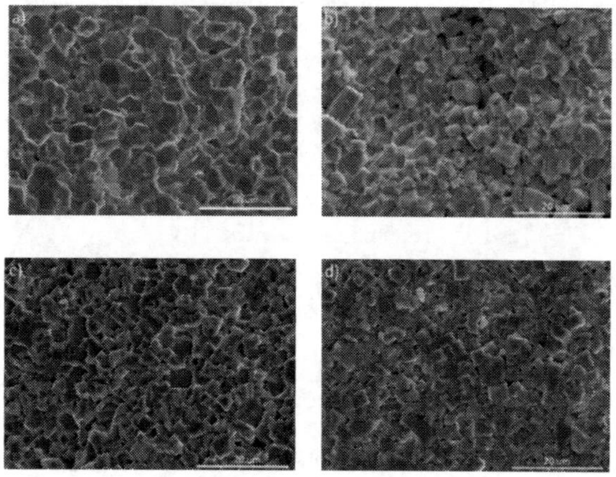

Fig.6 SEM photographs of fracture for a) $BiFeO_3$; b) $Bi_{0,97}Nd_{0,03}FeO_3$; c) $Bi_{0,97}Gd_{0,03}FeO_3$; d) $Bi_{0,97}Dy_{0,03}FeO_3$ ceramics.

The distribution of all elements was investigated with Energy Dispersion X-ray spectrometer (EDS) and carried out for randomly selected areas. Obtained results are very close to the calculated stoichiometric ratio for each prepared material. The small deviations from the theoretical composition have occurred but they do not exceed a value of 2.7%, what is consistent with the resolution of the utilized method of investigation.

Tab. 2 Theoretical and experimental content of element (calculation for simple oxide) for $Bi_{0.9}A_{0.1}FeO_3$ ceramic.

		BiNdFeO	BiGdFeO	BiDyFeO
Oxide content by EDS measurement [%]	Bi_2O_3	68,44	68,154	67,972
	A_2O_3	5,491	5,891	6,143
	Fe_2O_3	26,064	25,953	25,884
Theoretical content of oxides [%]	Bi_2O_3	68,02	67,9	67,95
	A_2O_3	5,5	6,05	6,21
	Fe_2O_3	26,48	26,05	25,84
Content error [%]	Bi_2O_3	0,614	0,373	0,032
	A_2O_3	0,164	2,699	1,091
	Fe_2O_3	1,596	0,374	0,17

4 Conclusion

By means of the mixed oxide method followed by pressureless sintering $Bi_{1-x}A_xFeO_3$ (A = Nd, Gd, Dy, for x = 0, 0.03, 0.05, 0.07, 0,1) ceramics was successfully fabricated from stoichiometric mixture of Bi_2O_3, Fe_2O_3 and $Nd_2O_3/Gd_2O_3/Dy_2O_3$ powders, via the solid state reaction route. The thermal analysis method enables to determine the optimum synthesis temperature. The temperature of synthesis was chosen as T=800°C. The density of ceramic samples depends on the type and content of dopant. Increase of the content of neodymium and dysprosium increases the density of the obtained materials, but with increasing of gadolinium dopant the density of the ceramics decreases. Analyzing the SEM image it can be noted, that addition of the individual dopants did not affect the packing of the beans. The shape of the grains changed. It was found that $Bi_{1-x}A_xFeO_3$ ceramics exhibited rhombohedral symmetry with R3c (36) space group.

References

[1] M. Fiebig, Revival of the magnetoelectric effect, J. Phys. D - Appl. Phys. **38**, R123 (2005)

[2] V.A. Khomchenko, V.V. Shvartsman, P. Borisov, W. Kleemann, D.A. Kiselev, I.K. Bdikin, J.M. Vieira, A.L. Kholkin, Effect of Gd substitution on the crystal structure and multiferroic properties of $BiFeO_3$, Acta Materialia **57**, 5137–5145 (2009)

[3] N. Hur, S. Park, P.A. Sharma, J.S. Ahn, S. Guha, S.-W. Cheong, Nature **429**, 392–395 (2004)

[4] V.Koval, I.Skorvanek, M.Reece, L.Mitoseriu, H.Yanc, Effect of dysprosium substitution on crystal structure and physical properties of multiferroic $BiFeO_3$ ceramics, Journal of the European Ceramic Society Volume **34**, Issue 3 (2014)

[5] D.G. Chena, X.G. Tangb, Q.X. Liuc, X.F. Chengd and Y. Zoue , The Nd-doping effect on dielectric abnormity of $BiFeO_3$ ceramics below the Néel temperature, Materials Science Forum Vol. **687**, pp 439-446 (2011)

[6] Kazhugasalamoorthy S., Jegatheesan P., Mohandoss R., Giridharan N.V., Karthikeyan B., *Investigations on the properties of pure and rare earth modified bismuth ferrite*, Ceramics, Journal of Alloys and Compounds **493**, 569–572 (2010)

[7] Pradhan S.K., Das J., Rout P.P., Das S.K., Mishra D.K., Sahu D.R., Pradhan A.K., Srinivasu V.V., Nayak B.B., Verma S., Roul B.K., Defect driven multiferroicityin Gd doped BiFeO3 at room temperaturę, Journal of Magnetism and Magnetic Materials **322** , 3614–3622 (2010)

[8] Sosnowska I., Peterlin-Neumaier T., and Steichele E., *Spiral magnetic- ordering in bismuth ferrite*, J. Phys. C: Solid State Phys. **15**, pp. 4835-4846. (1982)

[9] A. Gajovic, S. Sturm, B. J. War, A. Santic, J. K. Zagar, M. Ceh, The Synthesis of Pure-Phase Bismuth Ferrite in the Bi–Fe–O System Under Hydrothermal Conditions without a Mineralizer, J. Am. Ceram. Soc.,**93**[10] 3173–3179 (2010)

[10] M.S. Bernardo, T. Jardiel, M. Peiteado, A.C. Caballero, M. Villegas, *Reaction pathways in the solid state synthesis of multiferroic* $BiFeO_3$, Journal of the European Ceramic society, **31** (2011) 3047–3053

[11] Khesro A., Boston R., Sterianou I., Sinclair D.C., and Reaney I.M., Phase transitions, domain structure, and pseudosymmetry in La- and Ti-doped $BiFeO_3$, Journal Of Applied Physics **119**, 054101 (2016)

[12] N. A. Lomanova , V. V. Gusarov, Influence of synthesis temperature on $BiFeO_3$ nanoparticles formation, Nanosystems: Physics, Chemistry, Mathematics, **4** (5), P. 696–705 (2013)

[13] S.T. Zhang, Y. Zhang, M.H. Lu, C.L. Du, Y.F. Chen, Z.G. Liu, Y.Y. Zhu, N.B. Ming, Substitution-induced phase transition and enhanced multiferroic properties of $Bi_{1-x}La_xFeO_3$ ceramics, Appl. Phys. Lett. **88** (2006)

[14] V.F. Freitas, H.L.C. Grande, S.N. de Medeiros, I.A. Santos, L.F. Cotica, A.A. Coelho, Structural, microstructural and magnetic investigations in high-energy ball milled $BiFeO_3$ and $Bi_{0.95}Eu_{0.05}FeO_3$ powders, J. Alloy Comp. **461** (2008)

[15] Y.-Jie Zhang, H.-Guo Zhang, J.-Hua Yin, H.-Wei Zhang, J.-Ian Chen, W.-Quan Wang, G.-Heng Wu, Structural and magnetic properties in $Bi_{1-x}R_xFeO_3$ (x = 0-1, R = La, Nd, Sm, Eu and Tb) polycrystalline ceramics, J. Magn. Magn. Mater. **322** (2010)

[16] W. Kraus, G. Nolze, Powder cell - a program for the representation and manipulation of crystal structures and calculation of the resulting X-ray powder patterns, Journal of Applied Crystallography, **29**, 301-303 (1996)

[17] H.M. Rietveld, The Rietveld method – a historical perspective, Austr. J. Phys., 113–116 (1988)

[18] Z. Bojarski, E. Łagiewka, Rentgenowska analiza strukturalna, Wydawnictwo Uniwersytetu Śląskiego, Katowice (1995)

[19] T. Pikula, J. Dzik, A. Lisinska-Czekaj, D. Czekaj, E. Jartych, Structure and hyperfine interactions in $Bi_{1-x}Nd_xFeO_3$ solid solutions prepared by solid-state sintering, Journal of Alloys and Compounds **606**, 1–6 (2014)

Novel carbon fibers reinforced composites based on polysulfone matrix

Dilyus Chukov[1,a], Sarvarkhodzha Nematulloev[1], Andrey Stepashkin[1], Aleksey Maksimkin[1], Dmitriy Zherebtsov[1] and Victor Tcherdyntsev[1]

[1]*NUST "MISiS", Center for composite materials,119049 Leninskiy pr. 4, Moscow, Russia*

Abstract. The aim of this study is to create composites based on the high-temperature polymer reinforced with the carbon fibers and to study interfacial interaction between carbon fibers and polymer matrix. We propose a new method to obtain polysulfone based composite materials reinforced with high-modulus carbon fibers. The influences of thermal oxidation of carbon fibers on mechanical and thermal properties of the composites were studied. It was found that the obtained composite materials have sufficiently high mechanical properties, tensile strength up to 1047 MPa and Young's modulus up to 70.9 GPa were found. Considerable interest to the polymer composites is associated with their high performance and good mechanical and thermal properties, which enable a broad range of aerospace, automotive and medical applications. Additionally, the manufacturing process of such composites can easily be optimized and automatized, furthermore, it is not time-consuming process in relation with thermosetting polymer based composites.

1 Introduction

The development of composites combining excellent mechanical, thermal and other performance properties is a world trend in modern materials science. In past decades, a lot of attention was paid to composites based on thermoplastic polymers reinforced with continuous fibers. The growing interest in the composites caused by many attractive advantages in comparison to the widely used thermoset composites. The advantages include such properties as high fracture toughness and ductility, ease of shape forming prior to consolidation, significantly faster and lower cost manufacturing, longer shelf life of raw material [1–3]. The thermoplastic prepregs can be stored at normal environmental conditions with infinite shelf life in contrast to thermoset prepregs, whose shelf life is mainly limited. The synthesis of thermoplastic composites does not involve complex chemical reactions, therefore they can be processed with no time-consuming curing reactions [4]. Another undoubted advantage of thermoplastic composites is their ability to be reshaped and recycled by remelting the polymer matrix [5]. Therefore, to expand the scope of the composites application the development of new improved experimental techniques for the synthesis of thermoplastic composites is required.

Polysulfone (PSU) is a high performance engineering thermoplastic, which is often used in various kind of application, including tribological, aircraft and constructional applications. Possessing a number of advantages, this polymer has some drawbacks such as high processing temperature (300-350 °C) and low adhesion to carbon fibers (CF). It is well known that the realization of fibers mechanical properties in composites mainly depends on the interfacial interaction between fillers and polymer matrix [6,7]. Using CF as effective fillers is restricted by their low adhesion to polymer due to the chemical inertness of CF surface. Various methods, such as thermal or chemical modification of fiber surface [8–10], functionalization of polymer matrix [11], plasma treatment [12–14], nanotubes application [15-17] and other [18] were applied to improve the interfacial interaction between polymers and CF. In our previous papers the effect of thermal and chemical oxidation of CF surface on the interfacial interaction between CF and ultra high molecular weight polyethylene was investigated [19,20]. This work is aimed to create novel composite materials based on polysulfone reinforced with carbon fibers using polymer solvent impregnation of CF. We present the results of structure and mechanical properties investigation of the obtained composites.

2 Materials and methods

Ultrason S2010 (Germany) PSU and twill weave fabrics 3K-1200-200 based on high-modulus carbon fibers were used as matrix and reinforces, respectively. A novel method to produce composite materials was proposed, the scheme of process is shown in Fig. 1. To ensure a uniform impregnation of carbon fibers the approach for obtaining a polymer solution was used. N-Methyl-2-pyrrolidone was used as a solvent for PSU, 20 wt.% PSU solution was prepared using a magnetic stirrer. The PSU

[a] Corresponding author: dil_chukov@mail.ru

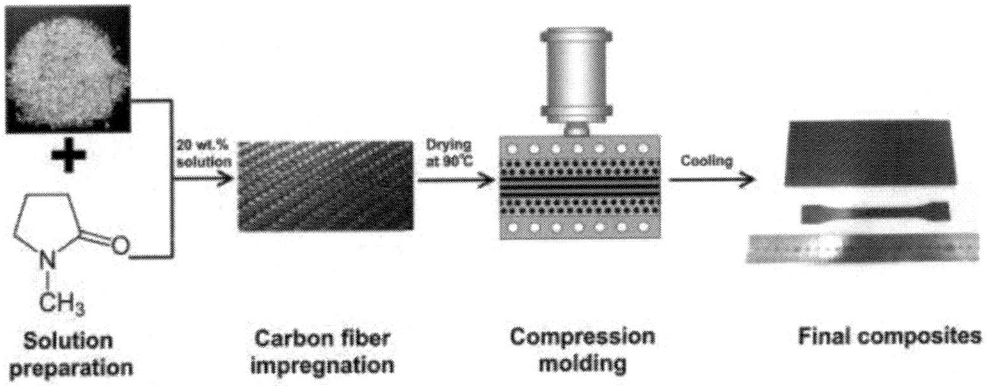

Figure 1. Preparation of polysulfone based composites

solution was used to impregnate the carbon fabrics. After impregnation the fabrics were dried at 90 °C. The fiber to polymer ratio of the composites was varied as follows: 50 wt% of the CF and 50 wt% of the PSU (denoted as 50/50), 60 wt% of the CF and 40 wt% of the PSU (60/40), 70 wt% of the CF and 30 wt% of the PSU (70/30).The prepregs obtained after drying were compression molded at 340 °C, applied pressure was of 10 MPa, and then cooled in air atmosphere. To improve the interfacial interaction between the components of the composites, surface modification of the CF by thermal oxidation at 300, 400 and 500 °C was applied.

The structure of the composites was studied using VEGA 3 TESCAN scanning electron microscope in a backscattered electron image mode. The mechanical tests were carried out using Zwick/Roell Z020 universal test machine equipped with 1 and 20 kN sensors and a contact strain measurement system MultiXtens. Tensile tests were carried out according to ISO 527:2009 for 110x10x2 mm samples at a speed of 10 mm/min. Flexural tests were carried out according to ISO 14125:1998 for 110x10x2 mm samples at a span length of 80 mm and at a speed of 10 mm/min. For all of the tests at least 5 samples were used. To study heat deflection temperature (HDT) Instron CEAST 6910 HDT/Vicat tester was used. The studies were carried out according to ISO 75 at a load of 1,8 MPa for 80x10x4 mm samples at a span length of 64 mm. The HDT measurements were carried out up to 1 mm deflection of the composites.

3 Results and discussion

Reinforcing fiber content is one of the main factors that determine the mechanical properties of fiber reinforced polymer composites. Stiffness, strength and conductivity, as well as other performance properties depend mostly on the fiber content. Figure 2a shows the results of tensile tests of produced composites with various fiber content, as well as composites reinforced with unidirectional fibers (UD). Obviously, for the UD composites the maximum values of the tensile strength (1220 MPa) and Young's modulus (73.9 GPa) are observed. The tensile strength and Young's modulus of composites reinforced with carbon fabrics is lower than that of UD composites and depend on CF content. The minimum value of tensile strength (748 MPa) and the Young's modulus (55 GPa) is observed for composites with 50% of the CF. An increase in the CF content up to 60% results in the tensile strength and the Young's modulus increase up to 880 MPa and 57.5 GPa, respectively. Meanwhile, the composites with 70% of the CF show lower values of the strength-elastic properties (775 MPa and 53.7 GPa, respectively). Three point flexural tests (Fig. 2b) show that the flexural strength behavior is similar to the tensile strength. The Young's modulus thus tends to increase, and from 56.7 GPa for the 50/50 composites this value increases up to 62.7 GPa, for the 70/30 composites.

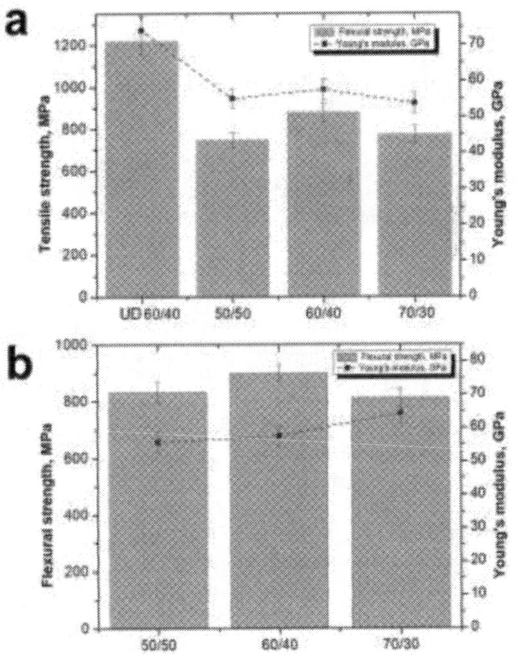

Figure 2. Mechanical properties of the composite materials: (a) tensile and (b) flexural tests

It can be concluded that at low CF content the strength of the composites is rather low because the presence of a large amount of the "weak" matrix phase. An increase in the fiber content results in a more dense fibers packing, while a decrease in the interlayer of polymer between the individual fibers occurs. Further increase in the CF content results in the decrease in the strength of the composites because in this case the matrix content is too low to binding the CF and to distributing the applied load to CF.

As it was mentioned earlier, the mechanical properties of composites mainly depend on the interfacial interaction between fibers and polymer matrix. To study the effect of thermal oxidation (TO) on mechanical properties the tensile and flexural tests of composites reinforced with fabrics with various TO temperatures were carried out. For these tests, the composites with an optimum fiber content of 60 wt.%, which showed the maximum values of strength-elastic properties, were used. Table 1 summarizes results of the studies.

Table 1. Mechanical properties of the composites reinforced with fabrics after surface modification

Designation	Tensile tests		Flexural tests	
	σ_{us}, MPa	E, GPa	σ_{us}, MPa	E, GPa
initial	948±21	56.2±3.8	899±28	57.6±2.4
300 °C	945±18	56.6±3.2	812±24	61.5±2.3
400 °C	967±25	61.6±3.6	882±26	66.2±2.1
500 °C	1047±16	70.9±4.2	1042±31	73.1±2.9

σ_{us} is ultimate strength; ***E*** *is Young's modulus*

An analysis of the results shows that the modification of the CF surface has a significant influence on the mechanical properties of the obtained composites. The maximum ultimate strength and Young's modulus values was obtained at TO temperature of 500 °C. The increase in mechanical properties of the composites is associated with the increase in interfacial interaction between the polymer and the reinforcing fibers, and a higher level of adhesion for composites reinforced with modified fibers results in improve in the strength-elastic properties of the composites. It is known that modification of CF can results in an increase in the concentration of oxygen on their surface, which means an increase in the quantity of oxygen-containing functional groups, such as the hydroxyl (- OH), carbonyl (-C=O) and carboxyl (-COOH) groups [21,22]. These functional groups contribute to the formation of the chemical bonds between CF and polymer, which results in the formation of strong interface in the composites.

Scanning electron microscopy showed (Fig. 3a) that the developed method for obtaining CF reinforced composites allows to well impregnation of the carbon fabrics with the polymer. The polymer matrix penetrates not only between individual threads of the fabric, but impregnates the threads, penetrating between the filaments of the carbon fibers.

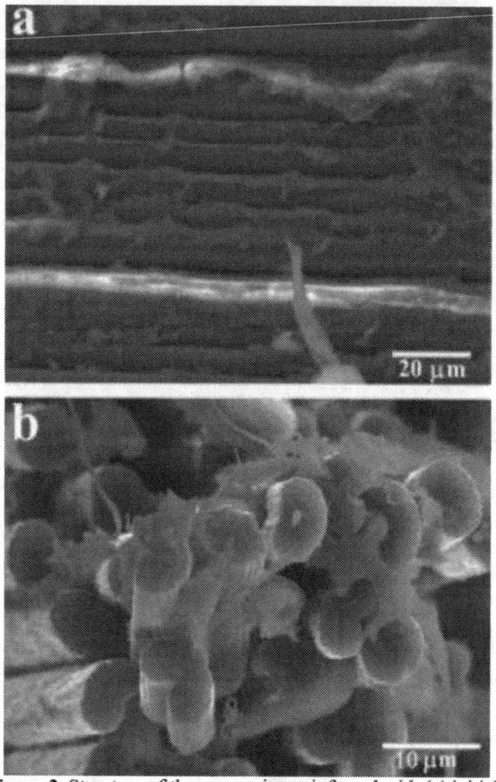

Figure 3. Structure of the composites reinforced with (a) initial and (b) oxidized at 500 °C carbon fibers

However, for a composites reinforced with the initial CF is typical a rather low level of interfacial interaction. The surface modification of the CF can improve interfacial interaction in the composites, and even after fracture, individual filaments can be coated with a thin continuous film of the polymer (Fig. 3b). The observed improvement in adhesion provides the high mechanical properties of composites reinforced with modified CF.

To evaluate the heat resistance a heat deflection temperature (HDT) of the produced composites were studied. The results of the studies are shown in Figure 4.

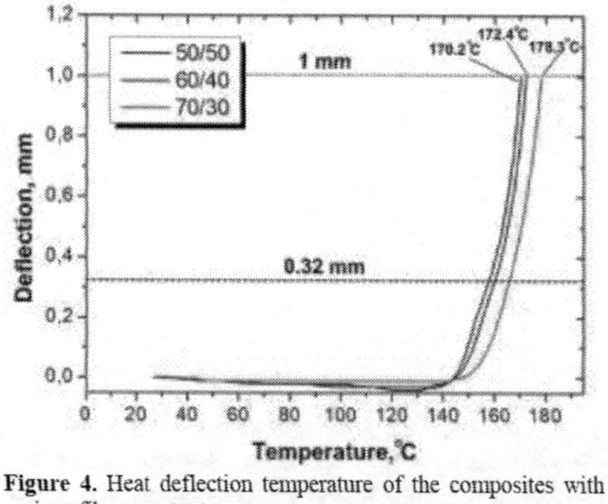

Figure 4. Heat deflection temperature of the composites with various fibers content

It was found that an increase in the fiber content is accompanied by an increase in the heat deflection temperature of the composites. The minimum HDT values (at deflection of 0.32 mm) are typical for a 50/50 composites (156.7 °C), and maximum HDT values were observed for a 70/30 composites (166.3 °C). Deflection of 1 mm is reaches at temperature range of 170-178 °C depending on CF content. Consequently, the dependencies obtained for composites with different polymer to fiber ratios can be interpreted as an increase in the heat resistance of the composites with increase in CF content. It is known that the loss of stability, including at the HDT tests, for amorphous polymers approximately coincides with glass transition temperature Tg values of the polymer. Consequently, loss of stability in the temperature range of 170-180 °C indicates that the heat resistance of the produced composites is mainly determine by the transition of the polymer matrix from the glassy state to the highly viscous (rubbery) one.

4 Conclusions

A novel method to obtain composites based on polysulfone reinforced with carbon fibers was proposed. The results of the structure, mechanical and thermal properties investigation of the produced composites were presented. The influence of the CF content and surface modification of the CF on strength-elastic properties of the composites was analyzed. It was found that the obtained composite materials have a sufficiently high level of mechanical properties (tensile strength up to 1047 MPa and Young's modulus up to 70.9 GPa), which is comparable to such well-known and widely used thermoset materials as the carbon fiber reinforced epoxy matrix composites. Obtained composites can be competitive in the construction materials market due to their faster and lower cost manufacturing, excellent thermal and mechanical properties.

Acknowledgments

The reported study was funded by Russian Science Foundation grant No. 18-19-00744.

References

1. J. Wang, K.-X. Li, H.-W. He, J.-L. Wang, and G.-H. Sun, Colloids Surf. Physicochem. Eng. Asp. **377**, 330 (2011).
2. T. Zhao, G. Palardy, I. F. Villegas, C. Rans, M. Martinez, and R. Benedictus, Compos. Part B Eng. **112**, 224 (2017).
3. H. Shi, I. F. Villegas, and H. E. N. Bersee, Compos. Part Appl. Sci. Manuf. **55**, 1 (2013).
4. V. Benoit, L. Cédric, and C. Alexis, Mater. Des. **63**, 56 (2014).
5. L. Ye, Z.-R. Chen, M. Lu, and M. Hou, Compos. Part Appl. Sci. Manuf. **36**, 915 (2005).
6. M. Sharma, S. Gao, E. Mäder, H. Sharma, L. Y. Wei, and J. Bijwe, Compos. Sci. Technol. **102**, 35 (2014).
7. J. Moosburger-Will, J. Jäger, J. Strauch, M. Bauer, S. Strobl, F. F. Linscheid, and S. Horn, Compos. Interfaces **24**, 691 (2017).
8. F. Severini, L. Formaro, M. Pegoraro, and L. Posca, Carbon **40**, 735 (2002).
9. C. M. Vaz, R. L. Reis, and A. M. Cunha, Biomaterials **23**, 629 (2002).
10. A. A. Stepashkin, D. I. Chukov, V. V. Cherdyntsev, and S. D. Kaloshkin, Inorg. Mater. Appl. Res. **5**, 22 (2014).
11. H. J. Salavagione and G. Martínez, Macromolecules **44**, 2685 (2011).
12. L. L. G. Silva, A. L. Santos, P. A. P. Nascente, and K. G. Kostov, J. Phys. Conf. Ser. **511**, 012082 (2014).
13. Y. Zhao, C. Zhang, X. Shao, Y. Wang, and Y. Qiu, J. Adhes. Sci. Technol. **25**, 2897 (2011).
14. K. Y. Rhee, S. J. Park, D. Hui, and Y. Qiu, Compos. Part B Eng. **43**, 2395 (2012).
15. A. R. Alian, S. I. Kundalwal, and S. A. Meguid, Compos. Struct. **131**, 545 (2015).
16. S. I. Kundalwal and S. A. Meguid, Eur. J. Mech. - ASolids **64**, 69 (2017).
17. S. I. Kundalwal and M. C. Ray, J. Mech. Mater. Struct. **9**, 1 (2014).
18. S. Tiwari and J. Bijwe, Procedia Technol. **14**, 505 (2014).
19. D. I. Chukov, A. A. Stepashkin, V. V. Tcherdyntsev, S. D. Kaloshkin, and V. D. Danilov, Inorg. Mater. Appl. Res. **5**, 386 (2014).
20. A. A. Stepashkin, D. I. Chukov, M. V. Gorshenkov, V. V. Tcherdyntsev, and S. D. Kaloshkin, J. Alloys Compd. **586**, S168 (2014).
21. J. P. Boudou, J. I. Paredes, A. Cuesta, A. Martínez-Alonso, and J. M. D. Tascón, Carbon **41**, 41 (2003).
22. R. Fu, L. Liu, W. Huang, and P. Sun, J. Appl. Polym. Sci. **87**, 2253 (2003).

MODELING OF THE FORMATION PROCESSES OF NANOCOMPOSITE COATINGS BY THE ELECTROCODEPOSITION

Alexander Vakhrushev[1,2], Eugene Molchanov[1] and Alexey Fedotov[1,2]

[1]*Institute of Mechanics, Udmurt Federal Research Center of the Ural Branch of the Russian Academy of Science, Izhevsk, Russia*
[2]*Kalashnikov Izhevsk State Technical University, Izhevsk, Russia*

Abstract. The coupled mathematical model of electrocodeposition process on the rotating cylinder electrode is presented. Low Reynolds k-e model with Abe-Kondoh-Nagano damping functions is used to describe mass transport of electrolyte. The cathodic and anodic processes are described by the tertiary current distribution theory. Mass transfer of electrolyte ions is described by diffusion-convection equation and is studied throughout the volume of the electrolyte cell. The simulation of electrocodeposition of Cu-Al2O3 nanoparticles on rotating cylinder electrode of are presented. It is found that the unsteady diffusion layer is formed close to the rotating electrode surface. Also it was found that the sign of the zeta potential of nanoparticles has a decisive influence on the possibility of forming the composite coatings. The results of mathematical modelling are in good agreement with the experimental data.

1 Introduction

The metal matrix composite electrochemical coatings (MMEC) with improved and unique operational characteristics, such as wear resistance, cracking resistance, anti-friction properties, corrosion resistance, radiation resistance and high adhesion to the substrate can be produced by electrocodeposition (ECD) technics [1-5].

The MMEC are prepared from the suspensions, representing electrolyte solutions with additives of certain quantity of a superfine powder. The particles are adsorbed onto cathode surface in combination with metal ions during ECD process and the metal matrix composite coating is formed. MMEC consists of galvanic metal (dispersion phase) and particles (dispersed phase).

The particles of synthetic diamond, ceramic or organic materials can be used as dispersed phase [4]. The particles of metals that can't be electrodeposited from aqueous electrolyte also can be used as dispersed phase. Usually, the particles with diameter from 4 to 800 nanometers are used in ECD process. The concentration of suspended particles in electrolyte bath usually ranges from 2 up to 200 g/l, which is producing composites coating with 10±1 vol. % of embedded particles [1-5].

The metals as Cu, Ni, Co, Cr, Zn, Ag, Fe, Au, As and their alloys can be employed as metal matrix of composite coating. The particles of oxides Al_2O_3, ZrO_2, TiO_2, SiO_2 and Cr_2O_3, carbides SiC, WC, TiC, various allotropic form C, nitrides Si_3N_4, polymers polystyrene, PTFE and several various materials can be used as dispersed phase.

The advantages of EPD process are the lowering of waste material in contrast with spraying or dipping technologies, the composite coating homogeneity of thickness and chemical composition even if the covered detail have a complex shape, the low-level of environment pollution [1]. The ECD process are not required the high operation temperature or pressure, vacuum or atmosphere of shielding gas. The ECD process realized at relatively low temperatures (in vicinity of ambient temperature), which is reduced to a minimum the undesirable chemical reactions or inter diffusion [5]. The thickness of coating can be accurately controlled by adjusting of applied current density. Nevertheless, there are several difficulties for example the agglomeration or sedimentation of nanoparticles and the nonuniformity on the nanoparticles concentration in the electrolyte.

The number of particles in the MMEC is a key factor which largely determines the properties of the composite coating. To produce the MMEC with specified physical and chemical properties it is necessary to define the main relation between the parameters of the ECD process and the properties of the MMEC. The determination of such empirical relationships, in addition to its purely practical significance, will allow developing the general production methods of the MMEC with required properties and the more accurately mathematical model.

It was found that there are several parameters that directly or indirectly effect on the properties of the MMEC. These parameters may be classified into three main categories (Table 1)

However, in practice it is difficult to reliably determine the influence of any single parameter on the particles content in the MMEC, since the influence of

[a] Corresponding author: Vakhrushev-a@yandex.ru

various parameters are often interconnected with each other.

Table 1. The system parameters.

Electrolyte parameters:	The particle parameters:	Parameters of electrodeposition:
•chemical composition; • pH value; • pollutants or additives; • life time of electrolyte.	• particle material; • particle size; • shape of the particles.	• concentration of the electrolyte; • current density; • electrolyte mixing; • temperature of the electrolyte.

Also, unfortunately, there are no experimental studies dedicated to determination the impact on the ECD process a series of above mentioned parameters.

2 ECD models

Historically the ECD process as an independent processing procedure was proposed in 1962 by Whithers [2]. Also, he made a first attempt to explain the mechanism of ECD process. He suggested that the particles suspended in the electrolyte solution may have a positive charge. Consequently, such particles are attracted to the negatively charged electrode (cathode) by electrophoretic forces.

In 1964 Martin and Williams [3] suggested that the particles move throughout electrolyte volume to the cathode surface due to the convective flow caused by agitation of the electrolyte. And then particles, located close to cathode surface, are mechanically entrapped by growing coating.

In 1967 Brands and Goldthorpe [4] suggested that van der Waals attraction force acts on the particles which were adsorbed on the electrode surface. This force keeps the particles on the surface during the time necessary for covering particles by metal matrix.

In 1972 Buzzard and Boden [5] suggested that the mechanism of ECD process can be divided into a series of sequential steps. On the first step the particles mass transfer to cathode surface occurs by the convection of electrolyte. After that the particles collides with the electrode surface and absorbs on it. Then the adsorbed particles are covered by the growing metal matrix.

In 1972 Guglielmi [6] proposed the mathematical model. The two successive steps were laid in the basis of this model. These steps were called weak and strong adsorption. The Guglielmi mechanism of MMEC formation is schematically shown in Figure 1.

During the first step there is only weak physical adsorption of the particles that have reached the electrode surface. On the surface of such particles there is a thin layer of adsorbed electrolyte ions which inhibits chemical reactions between the particles and the cathode surface.

The second step has an electrochemical nature. During this step the external electric field is applied to electrodes of electrochemical cell. The ions of the electric

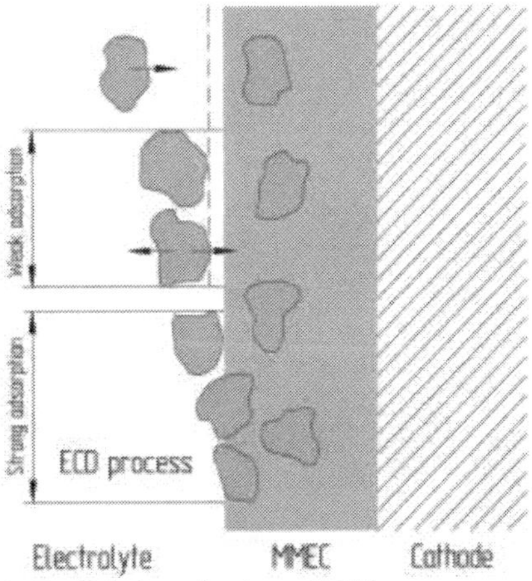

Figure 1. Guglielmi mechanism of ECD process.

double layer around the particles are reduced on the cathode by the influence of electric field and metal matrix is growing around weakly adsorbed particles.

Guglielmi model was experimentally verified on various electrochemical systems like Ni-SiC, Ni-TiO_2, Cu-Al_2O_3, Cr-C, Ag-Al_2O_3, Cu-SiC и Cu-P.

In 1974 Kariapper and Foster [7] based on experimental data suggested that the ions adsorbed on the surface of the particles have a dual role in the process. On the one hand, they define the zeta potential of particles and thus effect on particles electrophoretic mobility and on the other hand they physically links particles with cathode surface by ion reduction at the cathode. Kariapper and Foster divided the process mechanism on the two consecutive steps as well as Guglielmi. The first step is the physical adsorption, the second step has an electrochemical nature. On this step the ions adsorbed on the particle surface are reduced to metal atom. The mechanism of MMEC formation according to the Kariapper and Foster model is schematically shown in Figure 2.

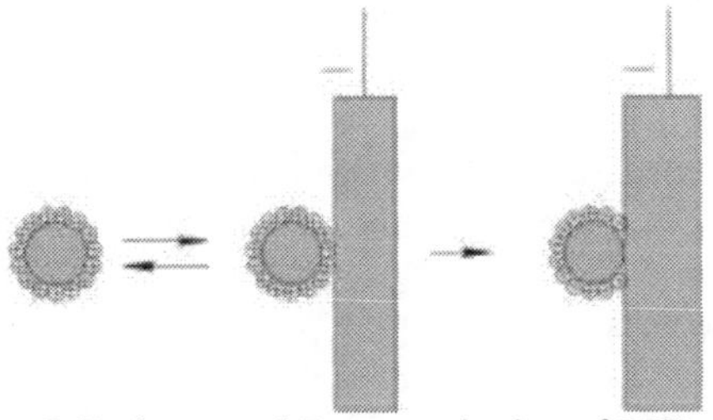

Figure 2. Kariapper and Foster mechanism of ECD process.

In 1983 Celis, Roos and Buelens [8] proposed the mathematical model that allows determining the amount of the embedded particles for electrochemical systems with a rotating disk electrode. Their model took into account the particle transport mechanism from the bulk electrolyte to the electrode surface. They mechanism of the ECD process is suggested to consist of 5 consecutive steps which are shown schematically in Figure 3.

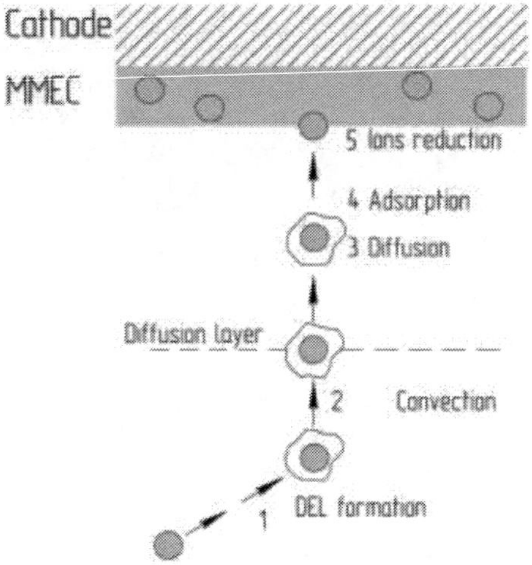

Figure 3. The scheme of the ECD process by Celisa.

On the first step electrolyte ions are adsorbed onto surfaces of particles suspended in the electrolyte. Therefore, the electric double layer is formed around each particle. On the second step the particles are transferred to Nernst diffusion layer due to the electrolyte convective flows caused by electrolyte mixing. Further, on the third step the particles reach the electrode surface through diffusion mass transfer. On the fourth step the particles surrounded by double electrical layer of electrolyte ions are weakly adsorbed on the electrode surface. On the last fifth stage the weakly adsorbed particles are covered by a layer of growing metal matrix due to reduction of the metal ions located near at the cathode surface and adsorbed at surface of particles.

In 1987 Valdes suggested the model [9], which takes into account the mass transfer of the particles in the electrolyte by diffusion, convection and migration. Valdes suggested that the intensity of particles adsorption process related to the intensity of reduction of electrolyte ions adsorbed on the particle surfaces.

In 1988 Guo [10] proposed the mathematical model allowed to describe the particles mass transfer by using the similarity criterions, which are dimensionless values. The Sherwood similarity criterion was upgraded by the Guo to describe the mass transfer process of particles:

$$Sh' = Co\,\mathrm{Re}^c\, Dm^d\, Sx^e\, Gq^f \quad (1)$$

where Sh' is the modified Sherwwood number, *Re* is the Reynolds number of hydrodynamic system, *Dm, Sx* and *Gq* are the dimensionless numbers that characterize the parameters of the particle electrical double layer, the particle concentration in the bath, the incorporation process of the particles in the metal matrix, *Co, c, d, e* and *f* are the empirical coefficients which are used to fit the model to the experimental data.

In 1991 Eng upgraded the Valdesa's model for a description of the ECD process on a rotating cylindrical electrode (RCE). In the Eng model the influence of the hydrodynamics was considered only by the thickness of the diffusion layer near the surface of the cathode, which is determined in accordance with the empirical expression:

$$\delta = 1.61 \cdot D^{1/3} \cdot \Omega^{-1/2} \upsilon^{1/6} \quad (2)$$

where δ is the thickness of diffusion layer, *D* is the diffusion coefficient, **Ω** is the angular velocity of rotating electrode and *v* is kinematic viscosity of electrolyte.

In 1992 Fransaer proposed mathematical model [11] to describe the ECD process on a rotating disc electrode of non-Brownian particles, i.e. particles with dispersion is less than 10^6 m^{-1}. This model is based on a trajectory analysis of particles, including convection and gravity, electrophoretic, Coulomb and inertial forces. The codeposition process was separated into two steps: the reduction of metal ions and the coincident deposition of particles. The probability of particles codeposition depends on the balance of forces acting on a particle near the electrode surface (Figure 4). Fransaer thought that if the interaction force between the particle and the surface of the cathode, $F_{adhesion}$, is more than the total shear force, F_{shift}, then particle codeposited into metal matrix otherwise particle leave the cathode surface.

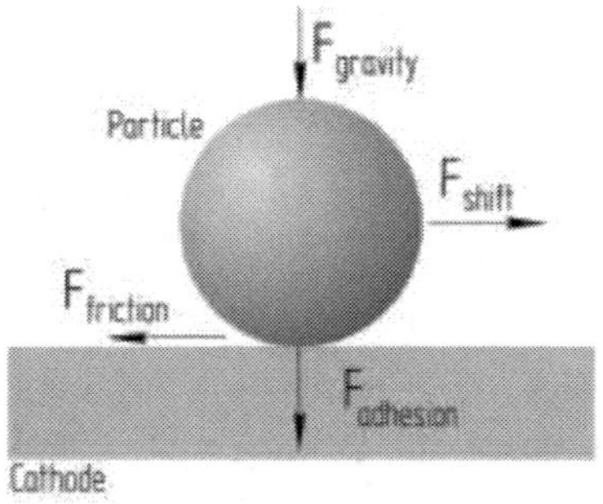

Figure 4. The forces acting on a particle near the electrode surface.

In 1993 Hwang and Hwang [12] proposed the more universal model than Celis model. They also suggested that the particle codeposition process is caused by reduction of electrolyte ions adsorbed on the particles. The intensity of which is determined by the parameters of the convective diffusion. In Hwang and Hwang model the mass transfer of particles and electrolyte ions are described by diffusion-convection equations and the calculated area is limited by the thickness of diffusion layer, within which the influence of convection is not considered.

In 2000 Vereecken proposed the mathematical model [13] to describe the ECD process of nanoparticles on a rotating disc electrode. The movement of the particles and the residence time with the electrode surface are considered in this model. The author investigated the mass transport of nanoparticles in the ECD process within the diffusion layer with convective diffusion equation. As in previous works Vereecken realize the mathematical simulation only within the Nernst diffusion layer.

In 2006 Huerta [14] proposed a mathematical model that describes the deposition of copper from sulfuric acid electrolyte with and without presence of chlorine anions and additives of polyethylene glycol nanoparticles. Author considered the adsorption process of polyethylene glycol nanoparticles and the reduction process of copper ions on the electrode as the multistep reactions during which intermediate compounds are formed. To describe this process author used the diffusion-convection equation and the equation of conservation of mass, as well as electroneutrality condition of the electrolyte. However, author performed the mathematical modeling within computational domain only at 3 times greater than thickness of the diffusion layer. Simulation was made in a one-dimensional scale. The influence of hydrodynamics on the mass transport of nanoparticles and electrolyte ion within computational domain was not accounted.

In 2007 Lee and Talbot [15] based on models of Celis and Vereecken developed a mathematical model describing the ECD process of alumina Al_2O_3 nanoparticles with copper matrix on a rotating cylindrical electrode. Authors have changed the basic model algorithm Vereecken to describe the ECD process at RCE. However, as in the previous models the computational domain was a narrow region near the cathode surface with the thickness of diffusion layer.

In 2013 Eroglu published a paper [16] on the modeling of ECD process of MMEC of Ni-SiC on a rotating cylindrical electrode. Eroglu proposed a new mechanism for description of nanoparticles adsorption. But as in previous models the effect of electrolyte hydrodynamics on mass transfer of nanoparticles and ions was taken into account only by the thickness of diffusion layer.

In 2015 [17,18] authors of this paper proposed a new mathematical model of ECD process. The proposed coupled mathematical model of ECD process on the rotating cylinder electrode is presented. The model takes into account the electrolyte turbulent flow. Low Reynolds k-e model with Abe-Kondoh-Nagano damping functions is used to describe mass transport of electrolyte. The cathodic and anodic processes are described based on the tertiary current distribution. Mass transfer of electrolyte ions is described by diffusion-convection equation and is studied throughout the volume of the electrolyte cell.

The results of hydrodynamic mathematical modeling of Cu-Al_2O_3 nanoparticles electrocodeposition on rotating cylinder electrode of are presented. For the first time ever, it is found that near the RCE surface the unsteady diffusion layer is formed by the electrolyte turbulent flow. Also, it is found that when the current density achieves their limit value the nonstationary diffusion layer with the variable thickness is formed near the electrode surface by the electrolyte turbulent flow.

The results of simulation were showed that the electrophoretic forces are considerably, and principal affected on nanoparticles content in composite coatings. Consequently, it is necessary not only takes they influence into account, but also utilized they influence on EPD process. The correct selection of the used electrolyte and surfactants can change the sign or value of the nanoparticles zeta potential.

The intensification of EPD process is occurred by changing parameters of hydrodynamic. The results of mathematical modeling are in good agreement with the experimental data.

3 Results of mathematical modeling

Simulation of copper electrodeposition with account of turbulent electrolyte flow was done at three studied rotational velocities (500, 1000, 1500 rpm).

The parameters of mathematical modeling corresponded to the experimental conditions [19]. While the simulation of electrocodeposition process of Cu-Al_2O_3 was done at three different concentrations of nanoparticles (39, 120 and 158 g/l) only at 1500 rpm. The computational time at all mathematical modeling was limited by 105 s. The voltage of electrochemical cell changes from the voltage of open circuit (+0.037mV vs SCE) to -0.5 V with a sweep rate of 5 mV/s during this time

The mathematical simulations of Cu electrodeposition confirmed that the hydrodynamics of electrolyte has a complex turbulent character at all studied rotation speed.

It is discovered that for all studied rotational speeds the concentration of Cu^{2+} ions close to surface of cathode is reduced to zero values when the cell voltage between the electrodes reaches -0.4 V relating to CRE. Therefore, at this cell voltage the applied current density is limited by the strong concentration polarization. Therefore, the diffusion mass transfer of Cu^{2+} ions to the cathode surface becomes the time-dependent step which determines the intensity of reaction of Cu^{2+} ions reduction at the cathode.

It is found for rotational velocity of RCE 500, 1000 and 1500 rpm that diffusion layer thicknesses are varied from 30 to 60, from 20 to 50, from 15 to 40 μm, respectively.

The rate of redox reaction achieves a limiting value because the diffusion mass transfer is limited. It should be noted that the oscillation of current density arise at all rotating speed of RCE when the diffusion mass transfer achieved a limiting value. The authors think that these oscillations can be explained by the turbulent flow near the cathode surface. The substitution by each other of zones with low (areas with blue color) or high speed (areas with red color) of electrolyte near the cathode surface leads to the appearance of current density oscillations (Figure 5). In more detail, the stagnant zones with small concentration of electrolyte ions are being appeared in some time point near to surface of cathode. Consequently, the diffusion layer with larger thickness is arisen too (Figure 6).

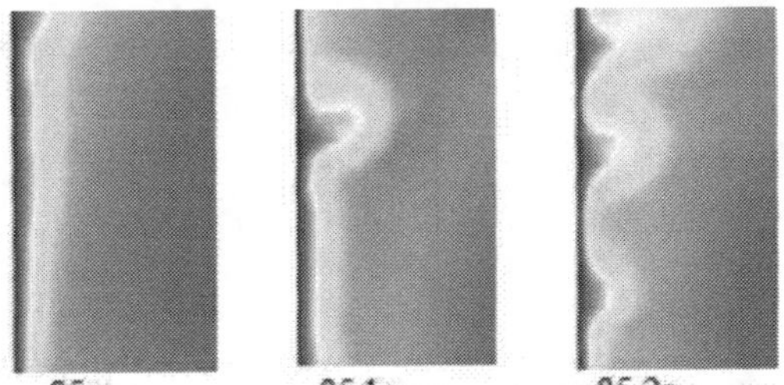

Figure 5. The substitution by each other of zones with low or high speed of electrolyte near the cathode surface

The simulation of ECD on the RCE was carried out. The results of modelling are presented on the figure 7. The results are point out that the flow of nanoparticles to the cathode linearly increases with increasing of applied current density. This means that the zeta potential of nanoparticles should be a positive.

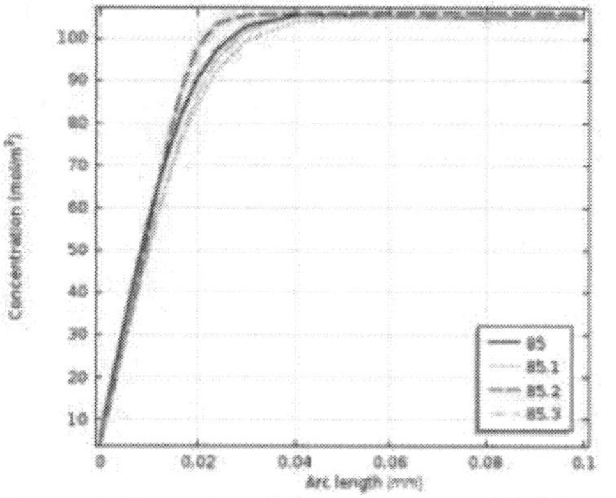

Figure 6. The various diffusion layers at 1000 RPM.

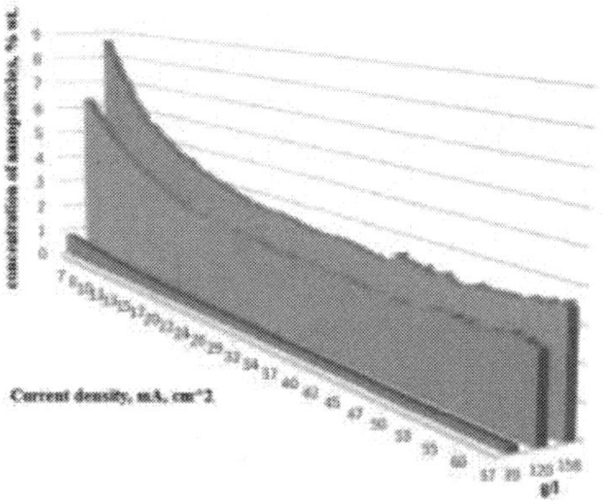

Figure 7. The concentration of Al_2O_3 particles in the composite coating.

4 Conclusions

The EPD process is the advanced method for using in practice the nanotechnology to creation new materials and coatings with unique working characteristics. The mathematical modeling of EPD process is the significant and needful step in the way of explanation of composite coatings formation mechanism.

The experimental determination of hydrodynamic optimal parameters is the complex and time-consuming process. Consequently, it is necessary to simulate how the electrolyte hydrodynamic affects on the studied process before the start to produce it at real conditions.

The coupled mathematical model of electrocodeposition process of Cu-Al2O3 nanoparticles on the rotating cylinder electrode is presented. The mathematical model account the mass transfer of electrolyte ions and nanoparticles, electrode reactions, the adsorptions of nanoparticles and turbulent flow of electrolyte. Near the surface of the electrode the zones with a low velocity of the electrolyte are formed. Which is significantly affected on the ions mass transfer to the electrodes surface.

It is found that the unsteady diffusion layer is formed near the electrode surface when the current density achieves the limited values.

The results of modeling confirm that the nanoparticles zeta-potential played the key role in the ability of composite coatings deposition. The sign of particles zeta-potential depends on the type of the crystal lattice of the particles (α or γ).

Next step of our works is the development the mathematical modeling of the ECD process using the methods of mesodynamics and molecular dynamics.

Acknowledgments

The investigation was carried with financial support from the Research Program of the Ural Branch of the Russian Academy of Sciences (project 18-10-1-29).

References

1. A. Gomes, I. Pereira, B Fernández., R. Pereir Advances in Nanocomposites - Synthesis, Characterization and Industrial Applications (Rijeka: InTech, 2011)
2. J.C. Whithers, Finishing, 26, 62 (1962)
3. R.V. Williams, P.W. Martin, Transactions of the Institute of Metal Finishing, **42**, 182 (1964)
4. E.A. Brandes, D. Goldthorpe, Metallurgic, **76**, 195 (1967)
5. R. Bazzard, P.J. Boden, Transactions of the Institute of Metal Finishing, **50**, 63 (1972)
6. N. Guglielmi, Journal of The Electrochemical Society., **119**, 137 (1972)
7. A.M. Kariapper, J. Foster, Trans. Inst. Met. Finish., **52**, 87 (1974)
8. C. Buelens, J.P. Celis, J.R. Roos, Journal of Applied Electrochemistry, **13**, 541 (1983)
9. J.L. Valdes, Journal of the Electrochemical Society., **134**, 223 (1987)
10. H. Guo, Q. Qin, A. Wang, Proceedings of the Electrochemical Society., **46**, 88 (1988)
11. J. Fransaer, Journal of The Electrochemical Society.,**139**, 413 (1992)
12. B.J. Hwang, J.R. Hwang, Journal of Applied Electrochemistry., **23**, 1056 (1993)
13. P.M. Vereecken, I. Shao, J. Electrochem. Soc., **147**, 2572 (2000)
14. M.E. Huerta, M.D. Pritzker, Journal of Electroanalytical Chemistry, **594**, 118 (2006)
15. J. Lee, J.B. Talbot, Journal of The Electrochemical Society, **154**, 72 (2007)
16. D. Eroglu, A.C. West, Journal of the Electrochemical Society, **160**, 354 (2013)
17. A. Vakhrushev, E. Molchanov, International Journal of Materials, **3**, 44 (2016)
18. A.V.Vakhrushev, E.K. Molchanov, Key Engineering Materials. **654**, 29 (2015).
19. J.L. Stojak, J.B. Talbot, J. Appl. Electroch. **31**, 559 (2001).

Research on the friction stir welding of Titanium Grade 1

R. Kosturek[1,a], M. Wachowski[1], T. Ślęzak[1], L. Śnieżek[1], J. Mierzyński[1], U. Sobczak[1]

[1] *Military University of Technology, Faculty of Mechanical Engineering, 2 gen. W. Urbanowicza str., 00-908 Warsaw, Poland*

Abstract. In this research the three different FSW joints of Titanium Grade 1 have been performed by using tool made of W25Re alloy with different welding velocity values. In order to investigate the influence of FSW process on microstructure of joined material the light microscope observations have been performed on the etched samples. It has been reported that significant grain refinement occurs in the stir zone in all analyzed samples. On the other hand, occurrence of weld defects, such as tunneling defect and kissing bond has been noticed. The microhardness analysis of the cross-section of the obtained joints indicates on microhardness increasing in the stir zone by 40-60 HV0.1 compared to the base material. Peak hardness of stir zone in the researched samples has tendency to decrease along with increasing of welding velocity. The strength of obtained joints was designated in the uniaxial tensile tests and confronted with strength of base material. Despite the occurrence of weld defects the established joints efficiency contains in range 92-94%. It has been stated that Young's modulus of Titanium Grade 1 FSW joints is 15-19% lower in comparison to the base material. At the same time, no significant influence of FSW on the ductility of material were observed.

1. Introduction

One of the most promising lightweight materials for used in modern engineering are titanium alloys [1,2]. Due to their high specific strength, corrosion resistance and high melting point titanium alloys have a great potential to increase their participation as construction materials in aircraft and aerospace industry [3-5]. One of the problems, which limit their wider use are obstacles concerned with fusion welding process, such as significant workpiece distortion and low quality of joints [6-8]. The technology, which has potential for replacement the fusion welding as the basic technology of joining titanium is the friction stir welding (FSW) [9,10]. FSW technique is the solid-state welding process, widely used for joining light alloys, such as aluminum and magnesium alloys. The process of joining is based on friction between workpieces and the rotating tool, which generates heat leading to plasticizing of material to be welded [9,11-13]. The movement of rotating tool along the edges of two workpieces causes mixing of plasticized material, and as the result the creation of joint between them [9,12]. Although the joining of aluminum alloys is easy to perform due to relatively low temperature of its plasticization, in case of titanium the higher welding temperature is required [9,12]. It can be provided by using more heat-generated welding parameters, which forces the adaptation of welding tool made of material able to work in high temperature environment, having excellent mechanical properties, as well as, outstanding wear resistance [13-15]. For this reason, the materials such as W-La alloys, WC-Co and polycrystalline cubic boron nitride (PcBN) and are used for the tool material in FSW of titanium alloys [16,17]. One of the recent proposed material with excellent wear properties, superior to WC-Co alloys is W25Re alloy [16]. In this investigation the self-designed tool made of tungsten-rhenium alloy (W25Re) has been used to perform the friction stir welding joints of Titanium Grade 1. The properties, as well as, exact geometry of designed tool have been presented in the previous investigation [18]. The main aim of this research is to investigate the possibility of obtaining high-quality joint of titanium grade 1 in FSW process by using tool made of W25Re alloy.

2. Experimental procedure

The joints of Titanium Grade 1 plates were successfully obtained using friction stir welding method. Joints were made using 250 x 1000 x 3.2 mm plate samples. The welded plates were held firmly using fixtures to prevent distortion during process. Before welding surfaces of plates were machined in order to negate oxide influence. Welding was performed using a fully-controlled FSW machine ESAB Legio 4UT. Welds were made along the longitudinal direction of the sheet (parallel to the rolling direction of plates). The welding tool used was designed by authors and made of W25Re and equipped in conical smooth pin diameter of 5-6 mm and 3 mm length. To obtain the joints three sets of welding parameters were used (Table 1). During the FSW,

[a] Corresponding author: robert.kosturek@wat.edu.pl

a 2° tilt, plunge depth of 3 mm and 20 kN axial force were applied to the FSW tool. Argon shielding gas was used for protection of the stir zone during welding due to high solubility of hydrogen, nitrogen and oxygen in hot titanium, what can lead to brittleness of obtained joints.

Table 1. Samples designation.

Shortcut	Description
Ti-75	Tool rotation speed: 250 rpm Welding speed: 75 mm/min
Ti-100	Tool rotation speed: 250 rpm Welding speed: 100 mm/min
Ti-200	Tool rotation speed: 250 rpm Welding speed: 200 mm/min

The welded joints were sectioned perpendicular to the welding direction where metallurgical examinations and hardness measurements were carried out. In order to investigate the joint microstructure, the samples were examined using light microscope (Olympus LEXT OLS 4100). The samples were mounted in resin, grinded with abrasive paper of 80, 320, 600, 1200 and 2400 gradations, polished using diamond paste of 1 μm gradation and then etched in mixture of mixture of HF (5 ml), HNO_3 (5 ml) and H_2O (70 ml) to revel the microstructure of titanium.

The Vickers microhardness measurements of the polished cross sections were taken at near mid-thickness across the welds by applying a load of 0.1 kg. Microhardness distributions were prepared for each sample.

The strength of obtained FSW joints of Ti gr. 1 was assessed during uniaxial tensile tests. The tests were performed using Instron 8802 hydraulic pulsator equipped with extensometer with a gage length of 50 mm. The dog-bone shaped samples had a cuboidal measuring section with dimensions of 30x3 and length of 75 mm. Tensile tests were carried out according to ISO 6892-1.

3. Results

3.1 Microstructure

The light microscope observations revealed the microstructure of obtained joints (Fig. 3). In all cases the joints have a microstructure specific for friction stir welded titanium, which consists of stir zone (SZ), thermo-mechanically affected zone (TMAZ) and base material (BM). Independently from the welding velocity the width of the stir zone in all cases is equal to 5.5 mm at the half of its height. On the other hand, the width of the thermo-mechanically affected zone decreases along with increasing of the welding velocity. It is equal to 2.2 mm, 2 mm and 1.5 mm at the half of the joint height for the samples Ti-75, Ti-100 and Ti-200 respectively. Microstructure of the stir zone is characterized by ultrafine equiaxial grains (Fig. 1).

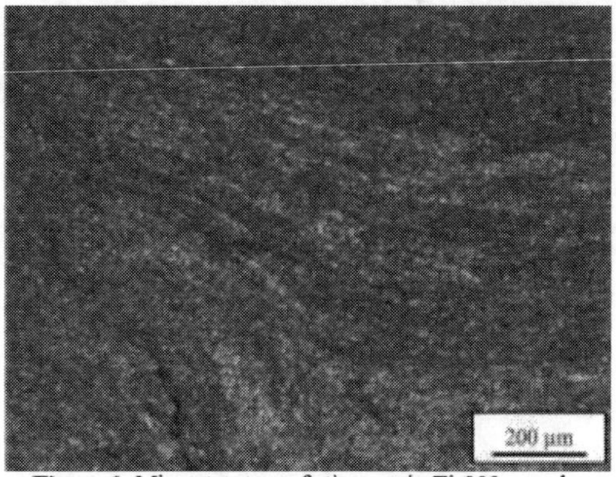

Figure 1. Microstructure of stir zone in Ti-200 sample.

This zone has been formed by direct acting of rotating tool on workpiece material. Severe plastic deformation and high temperature occurring in the stir zone during welding process results in formation of dynamically recrystallized ultrafine grain microstructure with average grain size of 5 μm. Between the stir zone and the base material occurs the thermo-mechanically affected zone (TMAZ) (Fig. 2).

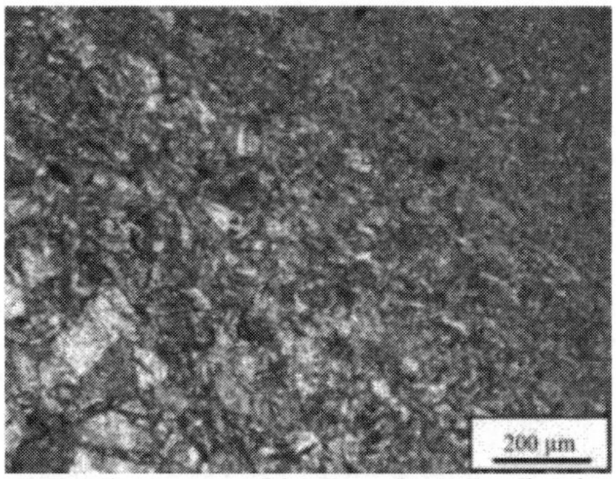

Figure 2. Microstructure of thermo-mechanically affected zone in Ti-200 sample.

TMAZ microstructure is characterized by thinner grains than base material, with visible deformation texture. In this area, the material has been subjected to the plastic deformation, as well as, has been influenced by high temperature, but not in sufficient way to cause the dynamic recrystallization of the microstructure. It has been noticed that the number of twin boundaries in this zone increased compared to the base material. The microstructure of TMAZ changes from the fine, fragmented grains to the grains elongated in the direction of material flow, and finally reaches the partially deformed grains zone. The obtained joint has two types of FSW defects: the tunneling defect, as well as, the kissing bond. The performed microhardness analysis allowed to estimate the influence of FSW process on hardening of material (Fig. 5).

3.2 Mechanical properties

 https://doi.org/10.1051/matecconf/201824201006

Three tests were conducted for the base material and each prepared FSW joint. Obtained results are presented in Fig. 4 in the form of stress-strain curves. In table 2 are showed the average values of Young's modulus E, yield strength $R_{0.2}$, ultimate strength R_m and elongation A.

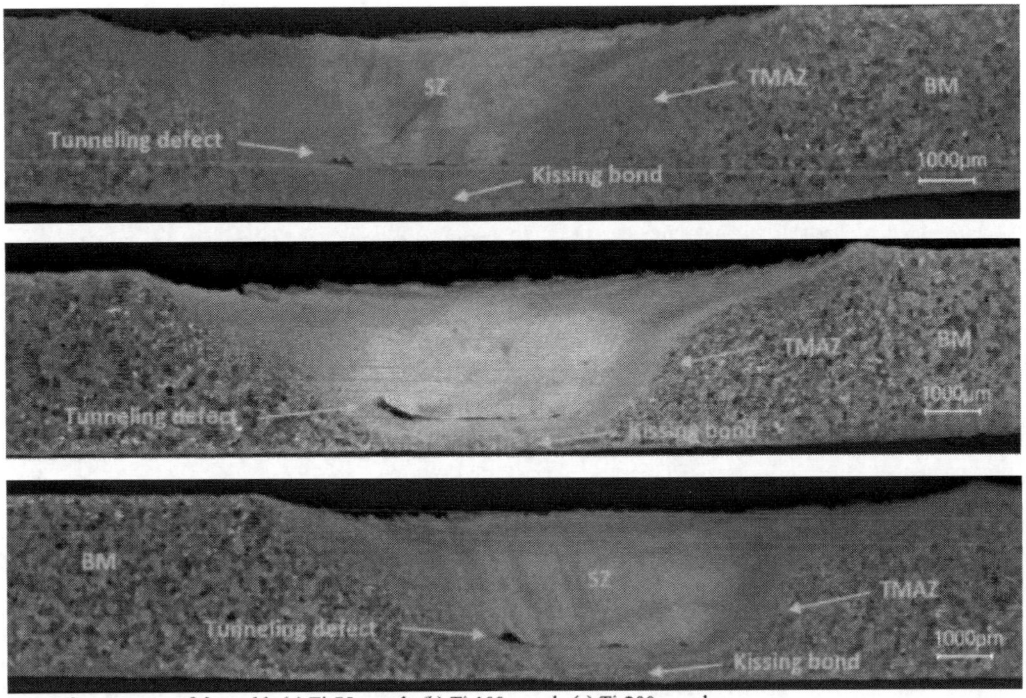

Figure 3. Microstructure of the welds (a) Ti-75 sample (b) Ti-100 sample (c) Ti-200 sample.

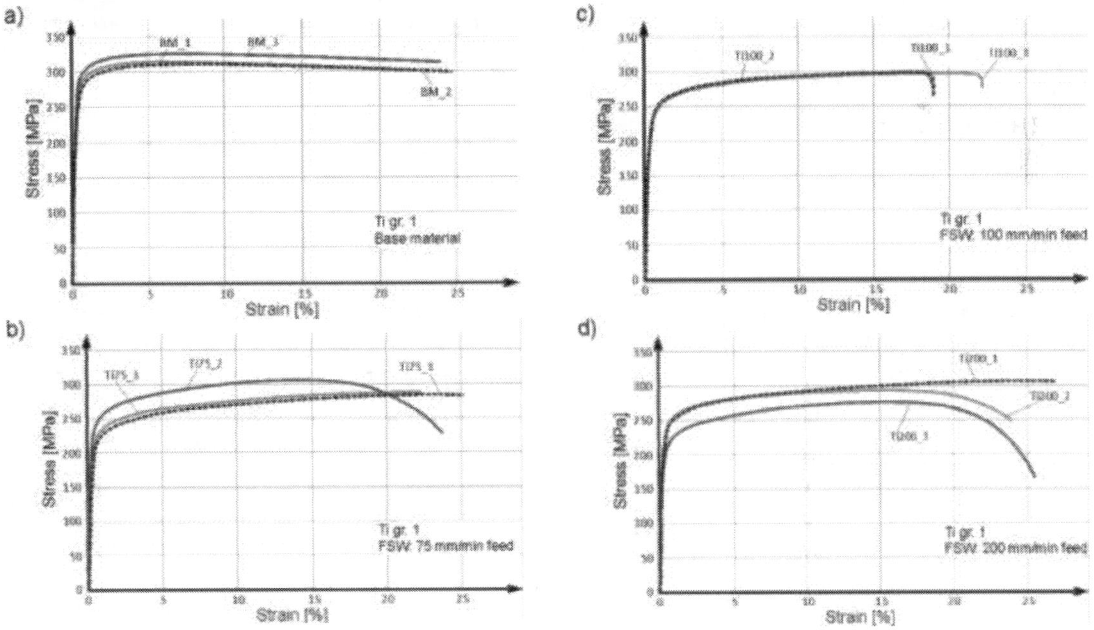

Figure 4. Stress-strain curves obtained during tensile tests performer on Ti gr. 1 base material (a) and samples with FSW joints produced under different feeds (b, c and d)

Table 2. Strength properties of Ti gr. 1 and FSW joints.

	BM	Ti-75	Ti-100	Ti-200
E [MPa]	$1.10 \cdot 10^5$	$0.91 \cdot 10^5$	$0.95 \cdot 10^5$	$0.93 \cdot 10^5$
$R_{0.2}$ [MPa]	266.8	214.5	221.7	215.3
R_m [MPa]	317.7	295.8	299.5	293.7
A_{50} [%]	23.3	23.3	19.7	24.9

It can be noticed the results obtained for base material and Ti-100 series are repeatable without more significant variations contrary to Ti-75 and Ti-200 series where every time one result is peculiar. Young's modulus E determined for FSW samples is 15-19% lower in comparison to base material what is influenced. In each case of joints yield strength $R_{0.2}$ and ultimate strength Rm are similar and lower than obtained for base material. On the other hand, the values of elongation for Ti-100 series are the lowest what means that these joints are more brittle compared to others and for Ti-200 series the elongation is the highest. Compared to the base material curves, the obtained joints have higher strain hardening coefficient.

3.6 Microhardness

It has been reported that significant grain refinement occurring in the stir zone causes the increasing of microhardness in this area up to the value of 180-200 HV0,1. The peak hardness always occurs in the stir zone and then gradually decreases across thermo-mechanically affected zone. The obtained microhardness values show a wide spread through cross-section of samples, what indicates on nonhomogeneous hardening of material in the friction stir welding process. The peak hardness of investigated samples decreases with the increasing of welding velocity.

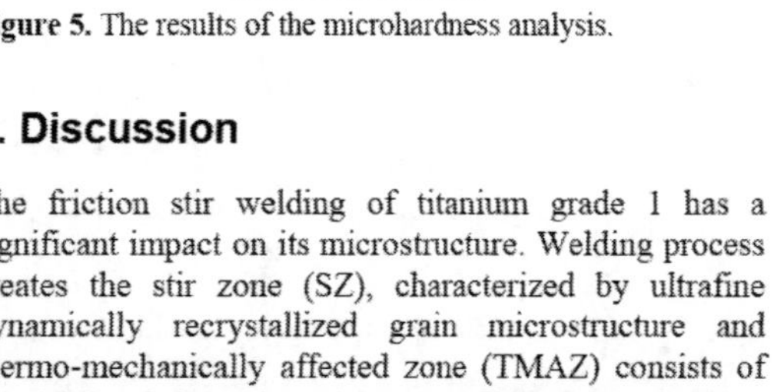

Figure 5. The results of the microhardness analysis.

4. Discussion

The friction stir welding of titanium grade 1 has a significant impact on its microstructure. Welding process creates the stir zone (SZ), characterized by ultrafine dynamically recrystallized grain microstructure and thermo-mechanically affected zone (TMAZ) consists of plastically deformed grains not affected by dynamic recrystallization phenomenon. In all samples has been reported the occurring of ultrafine SZ microstructure with grain size about 5μm. What is important, the peak temperature in friction stir welding process of titanium can exceeds the temperature of its allotropic transformation what can lead to formation of serrated grain boundaries in the stir zone, as the result of the $\beta \rightarrow \alpha$ transformation during cooling stage after formation of joint [11,19]. On the other hand, the allotropic transformation $\alpha \rightarrow \beta$, occurring in the heating stage of friction stir welding process and promoted by severe plastic deformation in the stir zone causes the dynamic restoration in this area [20,21]. No occurrence of serrated grain boundaries has been noticed in the stir zone, what suggest that the peak temperature during welding did not exceed the allotropic transformation temperature (882°C). The ultrafine grain structure of stir zone results in the hardening of material, what finds its confirmation in the results of microhardness analysis. The increasing of microhardness in the stir zone compared to the base material is equal to 65 HV0,1, 60 HV0,1 and 45 HV0,1 for welding velocities 75mm/min, 100mm/min and 200mm/min respectively. Despite of used welding velocity the width of the stir zone is equal to 5.5 mm at the half of its height. The grain size of thermo-mechanically affected zone microstructure slightly grows along with move from stir zone to base material (BM). That fact, together with observation of twin grain boundaries in the area closer to BM shows incomplete transformation of plastically defected grain microstructure by recrystallization process. Deformation texture of this zone reflects the material flow around the tool during welding process. In comparison to friction stir welding of aluminium alloys, in which the TMAZ is far wider, the titanium has significantly higher recrystallization rate, what results in formation of relatively thinner zone [22]. The microstructure observation together with results of microhardness analysis shows the relation between thickness of TMAZ and the used welding velocities. Together with increasing in welding velocity the TMAZ decreasing, what relates to the longer time of acting of the welding tool on the workpiece material. At the same time, the microhardness analysis results do not show occurrence of heat affected zone (HAZ) in the researched samples. In case of aluminium alloys, the HAZ can be characterized by local decrease in microhardness below the BM value. Softening of joint in this zone is causes by grain growth of undeformed material. The reason of HAZ absence in the titanium joints compared to the aluminium results from the lower thermal conductivity of titanium, as well as, its higher melting point, which determines the temperature of grain growth [23,24]. In the performed FSW joints, the two kinds of defects occurrence have been stated. On the bottom of joints cross-section the kissing bond defects is present. According to research, this imperfection is induced by insufficient degree of the stirring to breaking up the initial oxide layer on the joining surface [25,26]. The formation of kissing bond in this investigation could be caused by difference between the joined sheet thickness (3.2 mm) and pin's length (3 mm), what can lead to the lack of penetration and the ineffective stirring degree on the bottom of workpiece.

The second kind of defect present in performed joints is the tunnelling defect, which can be caused by both the improper welding parameters and the inappropriate welding tool shape. Considering the fact that this defect occurs in all three joints, regardless of used welding velocity, as the reason of improper material flow during welding process. Despite the imperfections the obtained joints exhibit very good mechanical properties. The average joint efficiency is equal to 93.1%, 94.3% and 92.4% for the samples Ti-75, Ti-100 and Ti-200 respectively. It is important to mention that the thickness of welded plates was about 3.2 mm and it had to be reduced to 3 mm by milling during the preparation of tensile test samples. This operation resulted in decreasing of the kissing bond defect's impact on the mechanical properties of tested joints. The results of tensile test indicate on almost negligible impact of tunnelling defect on the static mechanical properties of joints, what is an unusual issue. Good tensile strength of friction stir welded titanium is related to ultrafine grain microstructure, which occurs in the stir zone. On the other hand, no softening of material has been noticed in the thermo-mechanically affected zone. In all cases the yield strength of welded samples was lower than base material by about 40 MPa. It is worth noting that the welding process did not decrease the ductility of titanium grade 1 in most cases. It has been reported that Young's modulus of welded samples is 15-19% lower in comparison to the base material. This result can be explained by ultrafine grain structure of stir zone, as well as, the tunnelling defect, which has reduced slightly the real cross-section area of the tested samples.

5. Conclusions

In this study the Titanium Grade 1 sheets with 3.2mm in thickness have been welded by friction stir welding technique using different welding velocities and the tool made of W25Re. The microstructure, microhardness and mechanical properties of obtained joints have been investigated. It has been stated that despite the presence of weld imperfections, the tested joints exhibit a very good joint efficiency above 90% thanks to dynamically recrystallized ultrafine grain microstructure of the stir zone.

6. Acknowledgements

This work was supported by the university grant RMN 725/2017.

References

1. C. Veiga, J. Davim, A. Loureiro, Reviews on Advanced Materials Science **32**, 133-148 (2012)
2. C. Leyens, M. Peters, *Titanium and Titanium Alloys: Fundamentals and Applications* (2003)
3. M. Peters, J. Kumpfert, C.H. Ward, C. Leyens, Advanced Engineering Materials **5**, 419 – 427 (2003)
4. H.-X Jin, K.-X Wei, J.-M Li, J.-Y Zhou, W.-J. Peng, Zhongguo Youse Jinshu Xuebao/Chinese Journal of Nonferrous Metals **25**, 280-292 (2015)
5. E. Uhlmann, R. Kersting, T. Klein, & M. Fernando Cruz, A. Borille, Procedia CIRP **35**, 55-60 (2015)
6. M. Gurevich, *Problems of the welding of titanium and its alloys* (1970)
7. Y.M. Ahmed, K. Salleh, K. Sahari, M. Ishak, *Welding of Titanium (Ti-6Al-4V) Alloys: A Review* (2012)
8. K. Jata, P.R. Subramanian, A.P. Reynolds, T. Trapp, E. Helder, *Proceedings of the International Offshore and Polar Engineering Conference*, 22-27 (2004)
9. R. S. Mishra, M. W. Mahoney, *Friction Stir Welding and Processing, ASM International* (2007)
10. M. J. Russell, *10th World Conference on Titanium*, (2003)
11. J.-D. Kim, E.-G. Jin, S. Murugan, Y.-D. Park, Journal of Welding and Joining **35**, 6-15 (2017)
12. N. Xu, Q. Song, Y. Bao, Y. Jiang, J. Shen, X. Cao, Science and Technology of Welding and Joining **22**, 610-616 (2017)
13. G. Buffa, L. Fratini, F. Micari, L. Settineri, Transactions of the North American Manufacturing Research Institution of SME **40**, 785-794 (2012)
14. Q. Liu, R. Steel, J. Peterson, S. Horman, M. Collier, D.B. Marshall, J.B. Davis, M. Mahoney, *Proceedings of the International Offshore and Polar Engineering Conference* **4**, 298-302 (2010)
15. B. Thompson, *TMS Annual Meeting*, 105 – 112 (2011)
16. G. Buffa, L. Fratini, F. Micari, L. Settineri, Transactions of the North American Manufacturing Research Institution of SME **40**, 785-794 (2012)
17. Y. Zhang, Y. S. Sato, H. Kokawa, C. P. Hwan Seung. S. Hirano, Materials Science and Engineering: A **488**, 25-30 (2008)
18. U. Sobczak, M. Wachowski, L. Śnieżek, M. Bajkowski, K. Gocman, Machine Dynamics Research **41**, 53-64 (2018)
19. K. Gangwar, M. Ramulu, Materials & Design **141** (2017)
20. G. Buffa, A. Ducato, L. Fratini, Materials Science and Engineering: A **581**, 56-65 (2013)
21. Y.F. Sun, H. Fujii, Materials Science and Engineering: A **527**, 6879-6886 (2010)
22. G. Çam, S. Mistikoglu, Journal of Materials Engineering and Performance **23** (2014)
23. H. Fujii, H. Kato, K. Nakata, K. Nogi, Ceramic Transactions **198**, 51 – 56 (2011)
24. P. Podržaj, B. Jerman, D. Klobčar, Metalurgija **54**, 387-389 (2015)
25. N. Khan, A. Siddiquee, Z. Khan, S. Shihab, Journal of Alloys and Compounds **648** (2015)
26. K. P. Mehta, V. J. Badheka, Procedia Technology **23**, 513-518 (2016)

The effects of the heat treatment on the microstructure of Inconel 625/steel bimetal joint obtained by explosive welding

R. Kosturek[1,a], M. Wachowski[1], L. Śnieżek[1], M. Gloc[2], U. Sobczak[3,1]

[1] *Military University of Technology, Faculty of Mechanical Engineering, 2 gen. W. Urbanowicza Str., 00-908 Warsaw, Poland*
[2] *Warsaw University of Technology, Faculty of Materials Science and Engineering, 141 Woloska Str., 02-507 Warsaw, Poland*
[3] *The Building Research Institute, 1 Filtrowa Str., 00-611 Warsaw, Poland*

Abstract. In this investigation steel P355NH was successfully clad with Inconel 625 through the method of explosive welding. Explosively welded bimetal was subjected to the two separated heat treatment processes: stress relief annealing (at 620°C for 90 minutes) and normalizing (at 910°C for 30 minutes). In order to identify the microstructure of the joint and to investigate the influence of the heat treatment on it, the light and scanning electron microscope observations and microhardness analysis have been performed. In order to investigate the diffusion zone microstructure the scanning transmission electron microscope observation have been performed. It was stated that obtained joint has characteristic wavy-shape geometry with the presence of the melted zones and severe deformed grains of both joined materials. Strengthening of materials in joint zone was established with microhardness analysis. In both of the heat treatments the changes in the grain structure have been observed. The normalizing heat treatment has the most significant impact on the microstructure of the joint as well as the concentration of the chemical elements in the joint zone. It was reported that due to normalizing the diffusion zone has been formed together with precipitates in the joint zone.

1. Introduction

Corrosive wear represents a significant problem for the utilization of components of equipment operating in an aggressive environment, such as equipment for the chemical industry and geothermal power plants [1, 2]. The use of corrosion resistant alloys for the manufacture of components of equipment subject to wear is generally associated with considerable costs. An economical solution is to use clad materials where relatively inexpensive material, such as carbon steel, is covered by a layer of material that will provide protection from corrosion, e.g. acid resistant steel, which significantly reduces the costs of material used for the production of the equipment [3-5]. The present paper investigates steel P355NH clad with Inconel 625 as bimetallic material of the above type. Steel P355NH, unalloyed weldable constructional steel of fine grain structure, is used as material for the manufacture of pressure equipment operating under high temperature (up to 450°C) [6]. Since this steel is not corrosion resistant, its life cycle during operation in an aggressive environment may be prolonged by cladding it with a layer of Inconel 625, high-temperature creep resistant alloy of nickel and chromium with an additive of molybdenum and niobium that is characterised by high resistance to oxidizing and reduction environment as well as pitting and crevice corrosion and it also displays tolerance to a wide range of operating temperatures (from -150°C to 982°C) [7]. The manufacturing of clad material in which steel P355NH plays the load-bearing role and a layer of Inconel 625 alloy provides protection from corrosion may be achieved through explosive welding [8-11]. In the process, the energy released during detonation of the explosive is used for high velocity collision of metal materials [12-17]. The result of this high energy collision is bringing the surfaces of the welded materials close enough to each other to obtain interaction between their atoms and the consequent formation of a metallic bond between them [14, 18-20]. The considerable plastic deformation of the materials, produced as part of the process, significantly influences the structure of the materials and strengthens the joint by plastic deformation. Clad materials manufactured using this method are subjected to further technological processes to form specific equipment components for the industry, e.g. tube plates for heat exchangers and converter bottoms. In order to decrease the degree of hardening of the materials and reduce the residual stresses, the materials are subjected to heat treatment [21-23]. The annealing of bimetallic materials leads to significant structural changes within the joint, which may deteriorate the mechanical properties [24]. Steel clad with Inconel 625 using explosive welding has been the subject of numerous studies, which have confirmed that it is possible to obtain such a bimetallic joint through explosive welding [8-11]. However,

[a] Corresponding author: robert.kosturek@wat.edu.pl

literature does not contain sufficient investigation of the effect of heat treatment on structural changes in the joint. According to previous studies by the authors of this paper, normalizing of Inconel 625 – steel P355NH joint obtained by explosive welding deteriorates its resistance to shear by 33% (decrease from 572 MPa to 383 MPa, determined according to PN-EN13445:2014) [22]. The present paper aims to investigate the causes of such a drastic reduction in the resistance of the joint. Structural changes may involve both evolution of the grainy structure of the materials and diffusive transformations within the joint [21, 25-28]. Depending on the mutual solubility of the diffusing chemical elements of the alloy, brittle intermetallic compounds or new solid solutions may form in the joint area [29, 30]. The Kirkendall effect may accompany the diffusion changes, which consists in development of voids in the joint area due to differences in diffusion rates of individual alloying elements of the welded materials [25, 27, 31-33]. Additionally, Inconel alloys display a tendency to form precipitates (e.g. carbides, γ', γ'' and δ phases) upon long-term exposure to high temperature [34-37]. Importantly, plastic deformation of Inconel alloys exerts a considerable impact on the kinetics of precipitate formation, as it promotes their quicker development [38-40]. A factor promoting formation of carbides is heat treatment-induced diffusion of carbon from the steel to the Inconel alloy, which contains elements showing considerable affinity to carbon (Cr, Mo, Nb) [41]. For this reason, the layer of Inconel 625 alloy in the area of the joint produced through explosive welding is a zone prone to potential formation of precipitates during heat treatment of the bimetal.

2. Experimental procedure

The materials used in the study included a sheet of steel P355NH of 5 mm thickness and a sheet of Inconel 625 alloy of 3 mm thickness. The chemical composition of the materials is presented in Table 1 (Table 1). Samples were cut out of the sheets to investigate the microstructure of the materials at baseline, at the as-welded state. The process of explosive welding of steel P355NH and Inconel 625 alloy was performed by EXPLOMET High-Energy Techniques Works company. The explosive was modified ammonium nitrate/fuel oil (ANFO), with detonation velocity of 2700 m/s, determined using optical fibre sensors. Three samples were cut out of the bimetallic sheet produced by explosive welding, as shown on Fig. 1. One of the samples was subjected to heat treatment of stress relief annealing (at 620°C for 90 minutes), and another to normalizing (at 910°C for 30 minutes). The parameters of the heat treatment procedures were selected by EXPLOMET company based on their many years of experience in heat treatment of clad materials. As a result, three samples were obtained: a sample after explosive welding, one after stress relief annealing and one after normalizing (Table 2).

Table 1. Chemical composition of the alloys to be joined.

% Mass	Al	Cr	Fe	Mo	Nb	Ti	Ni
Inconel 625	0.16	21.5	4.6	8.7	3.32	0.18	Base
	C	Cr	Si	Mn	Ni	Cu	Fe
St. P355NH	0.18	0.02	0.35	1.19	0.22	0.2	Base

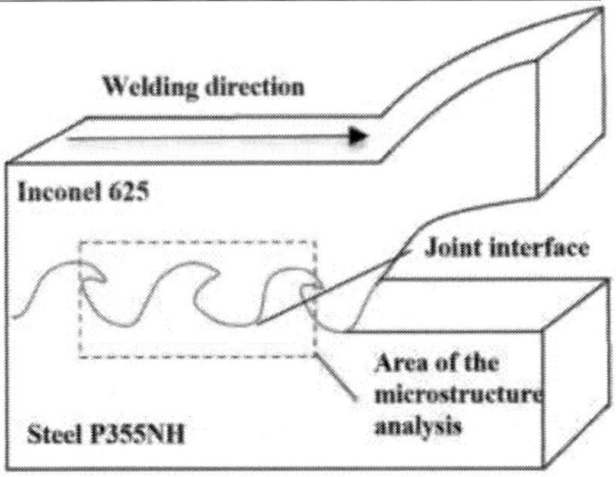

Figure 1. Area from which samples of bimetallic sheet were collected for microstructure investigation

Table 2. Samples designation.

Shortcut	Description
InSt EXW	Bimetallic joint Inconel 625/steel P355NH in the as-welded state
InSt HTR	Bimetallic joint Inconel 625/steel P355NH after stress-relief annealing (620°C / 90min)
InSt HTN	Bimetallic joint Inconel 625/steel P355NH after normalizing (910°C / 30min)

The samples were mounted in resin, grinded with abrasive paper of 80, 320, 600, 1200 and 2400 gradations and polished using diamond paste of 1 μm gradation. To reveal the microstructure of steel P355NH, 2% nital with etching time of 5-10 seconds was used and for Inconel 625 alloy, *acetic glyceregia* (15 ml HCl 38%, 10 ml of acetic acid 99%, 5 ml HNO_3 65%, 1-2 drops of glycerol) with etching time of 15 minutes. The microstructure of the samples was investigated using light microscope LEXT OLS 4100 and scanning electron microscope Joel JSM 6610. Additionally, an analysis of the microstructure of the joint zone was performed in the sample InSt HTN using scanning transmission electron microscope (STEM) Hitachi S-5500N. The samples for STEM were prepared using Dual beam system Hitachi NB-5000. The Vickers microhardness test was performed with loading of 100g. Microhardness distributions were prepared for each sample. The first two measurements were performed 200 μm from the joint line, in the layer of Inconel 625 alloy and in steel P355NH. Subsequently, measurement imprints were guided towards the edge of the samples, at the distance of 2000 μm.

3. Results

3.1 Microstructure of the raw materials

Microstructures of steel P355NH (Fig. 2a) and Inconel 625 alloy (Fig. 2b) in the as-welded state are presented on

Fig. 2. Steel P355NH shows ferrite-pearlite structure with banded distribution of the pearlite, typical of sheets after rolling. According to the measurements, it is characterised by the presence of fine equiaxial grains sized 15.5 ± 4.1 μm, specific to steel after normalizing. The microstructure of Inconel 625 alloy is more heterogeneous, with visible twins and grains sized 49.4 ± 15.6 μm. The measurement outcomes for microhardness in the as-welded state were 249.6± 16.3 HV0.1 for Inconel 625 alloy and 150.6 ± 4.9 HV0.1 for steel P355NH, respectively.

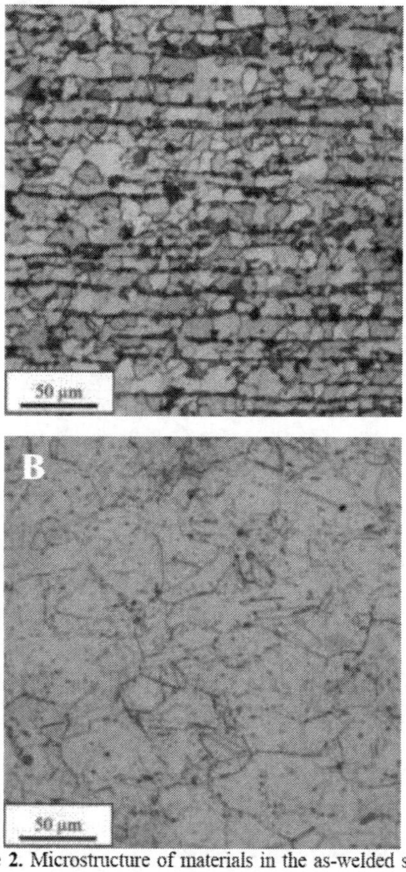

Figure 2. Microstructure of materials in the as-welded state a) steel P355NH b) Inconel 625.

3.2 Microstructure of InSt EXW joint

The joint produced by explosive welding displays a characteristic wavy geometry. The grains in the joint zone have undergone strong plastic deformation and have been observed to wrap to areas of intersurface waves in the samples after etching of steel P355NH (Fig. 3a) and Inconel 625 alloy (Fig. 3b).

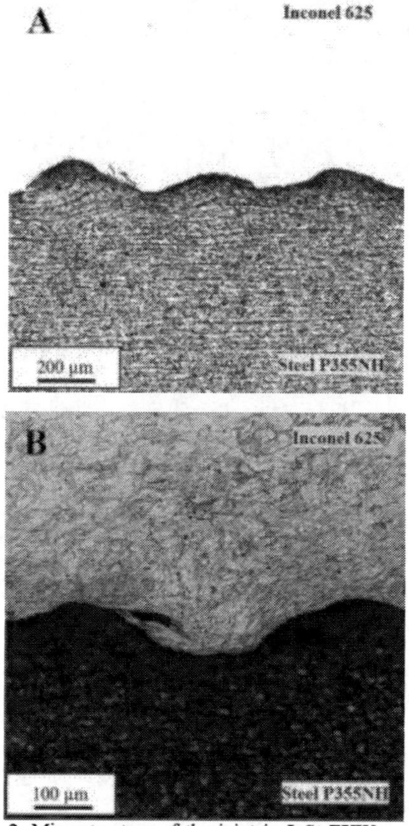

Figure 3. Microstructure of the joint in InSt EXW sample a) after etching of steel P355NH b) after etching of Inconel 625 alloy.

Additionally, fine, dynamically recrystallised grains of 1-2 μm size were found in steel P355NH close to the joint line (Fig. 4). The observation using scanning electron microscope showed melted zones, where the two welded materials blended (Fig. 5a). In the melted zones, the investigation revealed the presence of joint imperfections in the form of cracks and fragments of the surface layer of steel P355NH, which underwent partial fragmentation during welding, as evidenced by linear analysis of the chemical composition (Fig. 5b) and analysis of iron (Fig. 5c) and nickel (Fig. 5d) distribution at the sample surface (mapping).

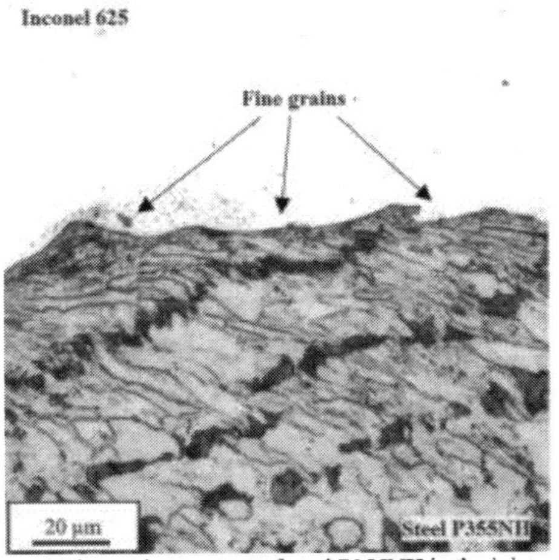

Figure 4. Fine grain structure of steel P355NH in the joint area.

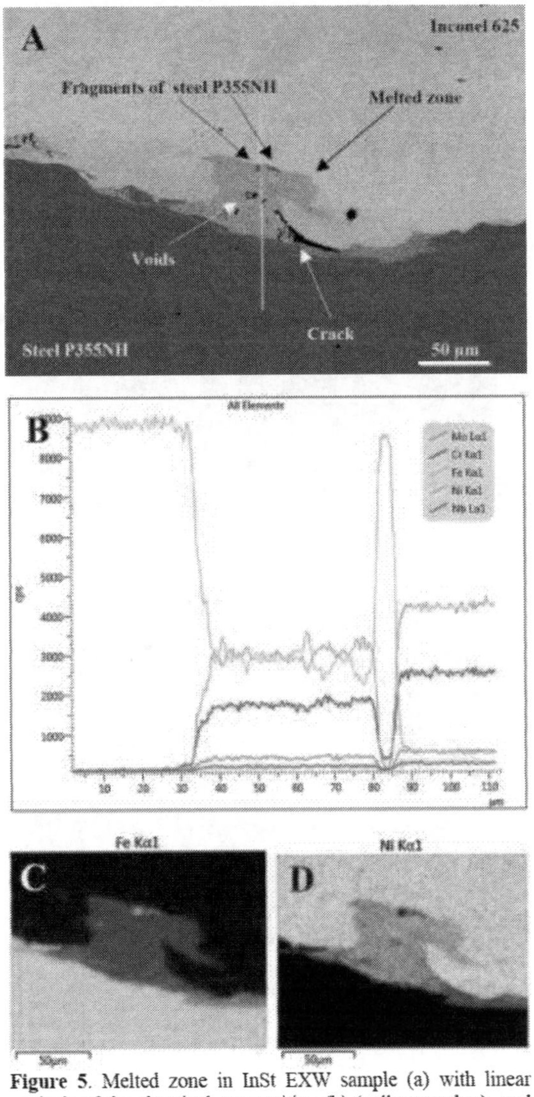

Figure 5. Melted zone in InSt EXW sample (a) with linear analysis of the chemical composition (b) (yellow marker) and mapping of iron (c) and nickel (d).

3.3 Microstructure of InSt HTR joint

Stress relief annealing, performed at 620°C for 90 minutes, led to partial recrystallisation of steel P355NH grains in the area of 20 – 30 µm from the joint, while the deformation texture of the material farther from the joint has been maintained (Fig. 6a). Observations of the microstructure of Inconel 625 alloy did not reveal visible changes in the grain morphology (Fig. 6b). Similarly, analysis of images from the scanning electron microscope did not show any visible changes in the concentration of alloying elements in the joint (Fig. 7).

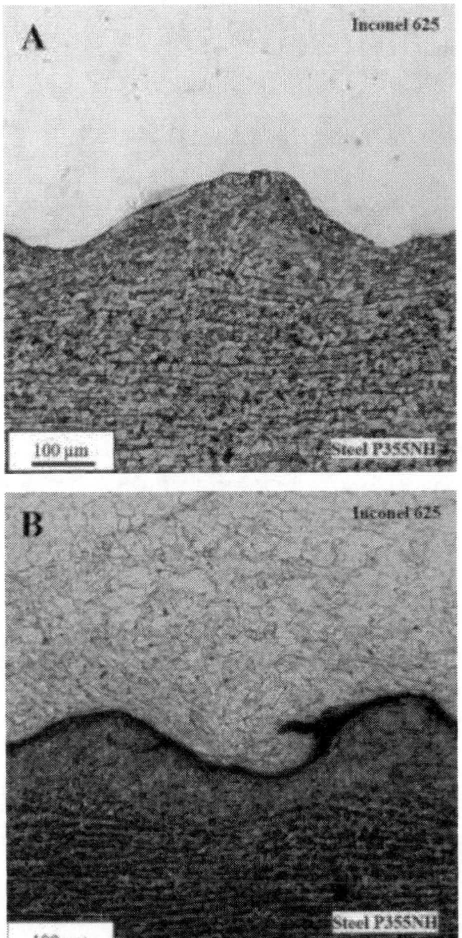

Figure 6. Microstructure of the joint in InSt HTR sample a) after etching of steel P355NH b) after etching of Inconel 625 alloy.

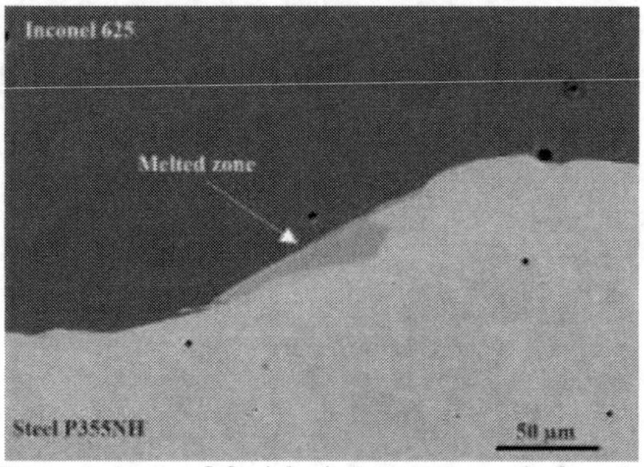

Figure 7. Image of the joint in InSt HTR sample from the scanning electron microscope.

3.4 Microstructure of InSt HTN joint

The microstructure of the joint subjected to normalizing at 910°C for 30 minutes changed considerably. Complete recrystallisation of the grainy structure of steel P355NH (Fig. 8a) and Inconel 625 (Fig. 8b) alloy was revealed, manifested by disappearing of the deformation texture and the presence of fine equiaxial grains in the joint. The observation using scanning electron microscope showed considerable changes in the concentration of alloying elements in the joint area. Alloying elements of Inconel 625 were found to diffuse into steel P355NH approximately 20µm-deep and precipitates, located mainly in the joint area, were revealed in Inconel 625 alloy (Fig. 9a). Diffusion of iron from steel P355NH into Inconel 625 alloy was found up to 5 µm deep. In the Inconel 625 alloy, *light* precipitates were revealed, suggesting a high concentration of alloying elements heavier than nickel, and *dark* precipitates, characterised by a high concentration of alloying elements lighter than nickel. Additionally, voids were found in the diffusion area (Fig 9b). The results of an analysis of alloying elements distribution on the surface of the sample in the joint area indicate an increased concentration of chromium in *dark* precipitates (Fig. 10b) and an elevated concentration of molybdenum in *light* precipitates (Fig. 10c). The results of nickel (Fig. 10a) and iron (Fig. 10d) distributions in the joint allow to observe a diffusion zone.

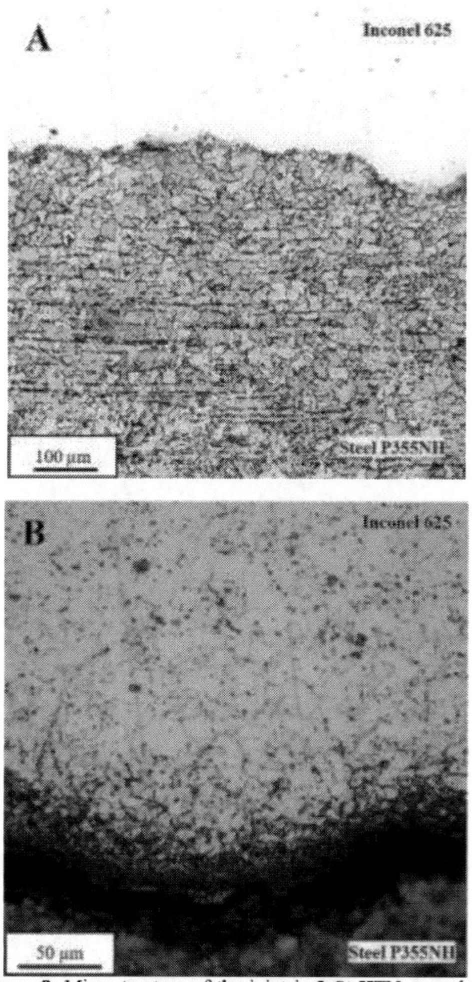

Figure 8. Microstructure of the joint in InSt HTN sample a) after etching of steel P355NH b) after etching of Inconel 625 alloy

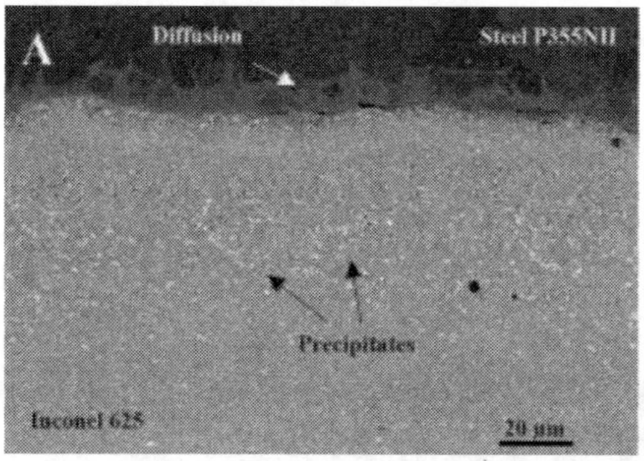

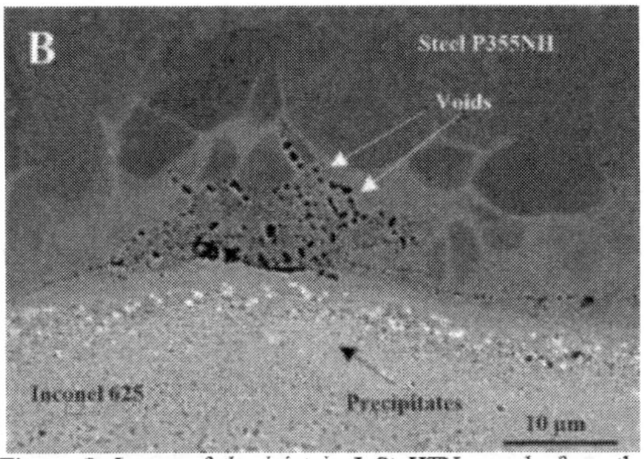

Figure 9. Image of the joint in InSt HTN sample from the scanning electron microscope a) precipitates in Inconel 625 alloy, b) diffusion zone

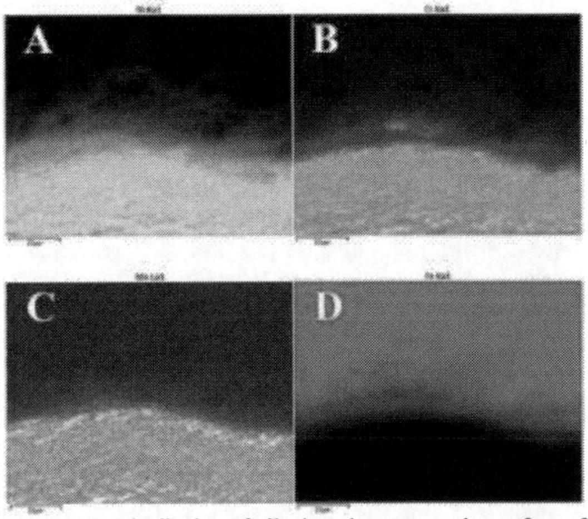

Figure 10. Distribution of alloying elements on the surface of InSt HTN sample in the joint area: a) nickel, b) chromium, c) molybdenum and d) iron.

3.5 Scanning transmission electron microscope observations of the diffusion zone

Observations of the diffusion zone performed on scanning transmission electron microscope indicates that this area consists of equiaxed grains (from side of Inconel 625 alloy) and columnar grains (from side of steel P355NH) (Fig. 11).

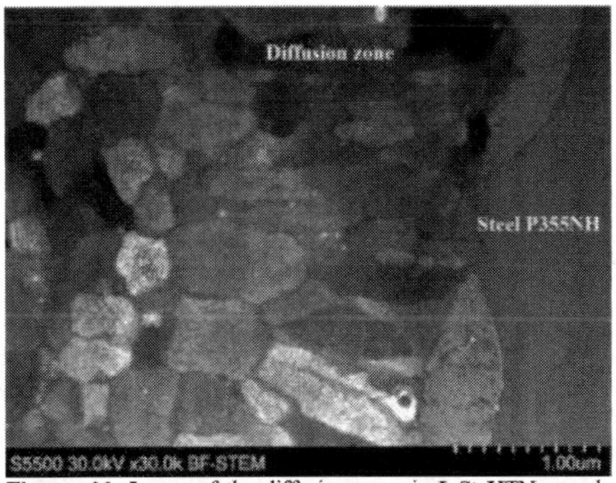

Figure. 11. Image of the diffusion zone in InSt HTN sample from the scanning transmission electron microscope.

Additionally, it has been reported the occurrence of precipitates with their width about 50 nm in the diffusion zone (Fig. 12).

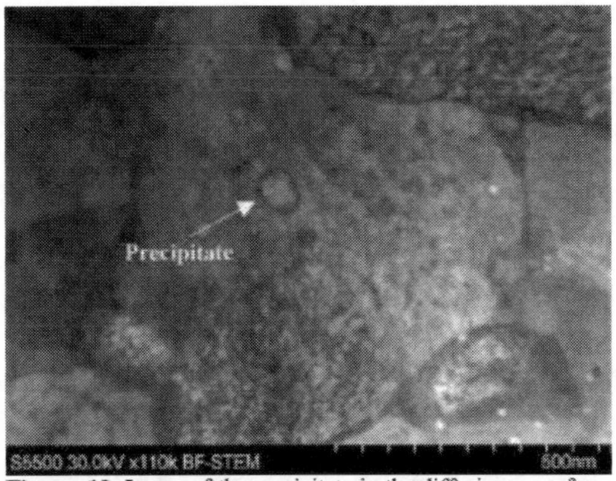

Figure. 12. Image of the precipitate in the diffusion zone from the scanning transmission electron microscope.

The results of linear analysis of the chemical composition of columnar grains (Fig. 13) and the precipitate (Fig. 14) in the diffusion zone are shown below.

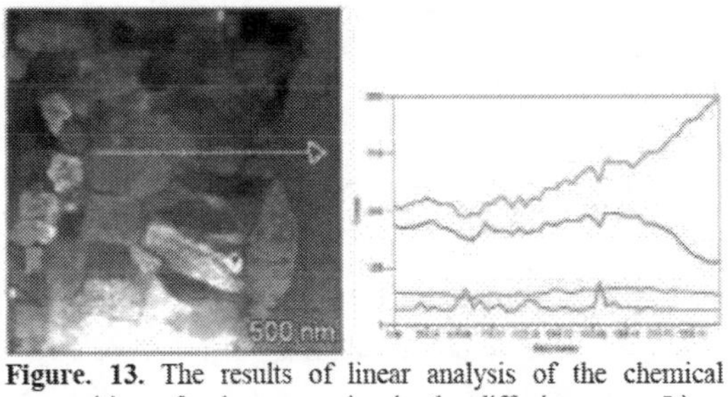

Figure. 13. The results of linear analysis of the chemical composition of columnar grains in the diffusion zone. Lines designation: Fe (green), Ni (blue), Cr (red), Nb (purple).

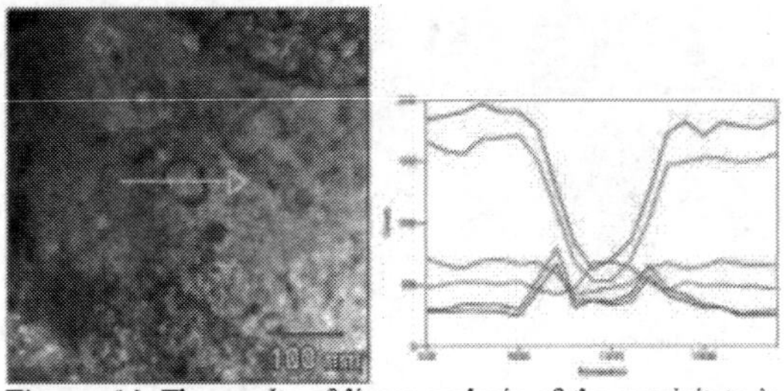

Figure. 14. The results of linear analysis of the precipitate in the diffusion zone. Lines designation: Fe (blue), Ni (purple), Cr (green), Nb (black), Mo (burgundy).

3.6 Microhardness analysis

The distributions of microhardness, prepared as part of the investigation, enable characterizing the degree of hardening of welded materials. The highest degree of hardening was revealed at the joint line, where materials underwent the strongest deformation. In Inconel 625 alloy, microhardness increased by ca. 200 HV and in steel P355NH by ca. 100 HV. A noteworthy finding is that hardening of steel P355NH in the joint shows a homogeneous nature, while in Inconel 625 alloy microhardness decreased with increasing distance from the joint. Stress relief annealing reduced the microhardness of steel P355NH by ca. 40 HV due to recrystallisation of grains and led to a monotonic decrease in the microhardness of Inconel 625 alloy. Heat treatment in the form of normalizing reduced the microhardness of steel P355NH to its baseline value, recorded before explosive welding.

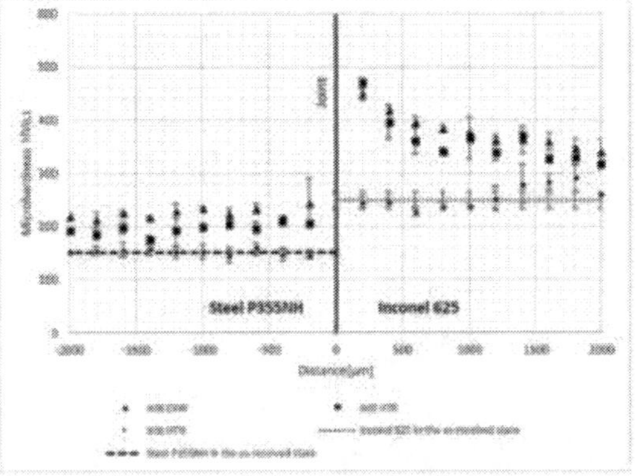

Figure 15. The results of microhardness analysis.

4. Discussion

The technique of explosive welding has produced a good quality joint between steel P355NH and Inconel 625 alloy, with characteristic wavy geometry and small amount of imperfections in the form of cracks, pores and melted zones. The melted zones have been found to display an evenly distributed blending of the welded materials, an increased concentration of imperfections and presence of fragments of steel P355NH surface layer. Light microscope observation of the bimetal microstructure after etching of steel P355NH and Inconel 625 alloy has revealed strong hardening of the materials in the joint area. Elongated compressed grains wrapping into intersurface waves have been found. Plastic deformation of the welded materials has been also reflected in the results of microhardness analysis, which suggests considerable hardening of steel P355NH and Inconel 625 alloy in the joint area. The microstructure of steel P355NH has shown fine, dynamically recrystallized grains close to the joint line. Heat treatment, both in the form of stress relief annealing and normalizing, has resulted in recrystallisation of steel P355NH grains in the joint area. The steel grains are finest at the joint line, which indicates cumulation of the strongest deformation in this area due to interaction of the welded materials. This finding is further confirmed by the analysis of heat-activated phenomena as a function of the temperatures used for heat treatment. As a result of stress relief annealing at 620°C, recrystallisation was found only in the joint area, while the banded grainy structure of the steel at a distance of ca. 100μm from the joint continued to show the original banded morphology. On the other hand, an increase of annealing temperature to 910°C results in a complete restructure of the grainy structure of steel P355NH, which had been defected through plastic deformation. It is a well known fact that the higher degree of plastic deformation, the lower energy is necessary to trigger and complete heat-activated phenomena and therefore the recrystallisation temperature is lower, which explains the incomplete restructure of the deformed steel structure after stress relief annealing. Changes in the concentration of alloying elements in the joint area after heat treatment in the form of normalizing are manifested through formation of both a diffusion zone and precipitates in Inconel 625 alloy. An analysis of images of the diffusion zone by scanning electron microscope leads to the conclusion that the diffusion of alloying elements of Inconel 625 alloy into steel P355NH takes place along the grain boundaries. Moreover, voids have been found in the diffusion zone, which are most probably due to the Kirkendall effect. An analysis of the phase equilibrium of iron-nickel may suggest that the diffusion zone is composed predominantly of nickel austenite γ and a small amount of nickel ferrite α (Fig.16).

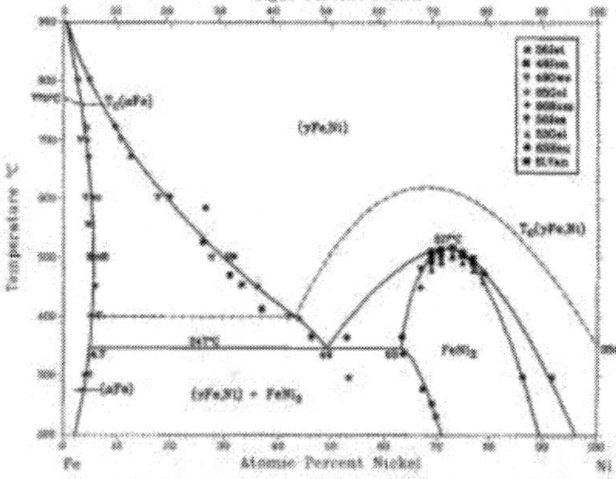

Figure 16. Graph of phase equilibrium for iron – nickel [42].

Precipitates due to normalizing in Inconel 625 alloy are located along the grain boundaries and twins, which has been revealed based on comparison of images of the sample after etching from the scanning electron microscope and light microscope. An analysis of an isothermal transformation diagram for Inconel 625 alloy leads to the conclusion that the parameters of the normalizing used are close to the area of carbide formation, mainly of MC and M_6C type (Fig. 17).

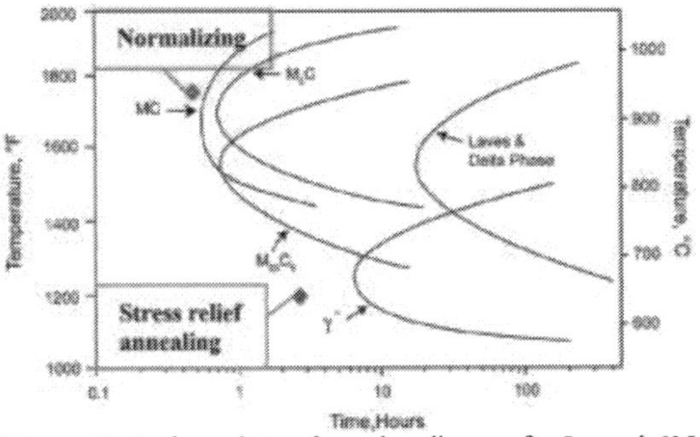

Figure 17. Isothermal transformation diagram for Inconel 625 alloy [35]. Values of stress relief annealing and normalizing have been marked

A noteworthy fact is that the strong plastic deformation of Inconel 625 alloy in the joint reduced the energy of heat-activated processes, including precipitate processes. An additional factor promoting formation of carbides is diffusion of alloying elements of steel P355NH into Inconel 625 alloy (in particular carbon, whose content in the steel is 0.18%). An analysis of distribution of alloying elements on the surface of the sample and microstructure observations using the scanning electron microscope enabled identifying two basic types of precipitates: *light* precipitates, rich in molybdenum, and *dark* precipitates, rich in chromium. Literature on diffusion in the bimetallic configuration of Inconel 625 alloy and low-alloyed steel precisely describes the precipitates as carbides of M_6C type in the case of high molybdenum content and of $M_{23}C_6$ type in the case of high chromium content [41]. The structural analysis of samples after normalizing indicates that both the diffusion zone and precipitates in Inconel 625 alloy represent the causes of deteriorated endurance of the joint compared to the as-welded state. In the diffusion zone, not only voids have been found, but also the structure itself, which contains soft nickel ferrite α on the side of steel P355NH and resistant nickel austenite γ on the side of Inconel 625 alloy, significantly deteriorates the quality of the joint. Carbides, forming along the grain boundaries and twins in Inconel 625 alloy, also reduce the coherence of the material in the most crucial zone, i.e. the joint. The results of scanning transmission electron microscope observations of the diffusion zone, allow to investigate the microstructure of the diffusion zone, which consist of equiaxed and columnar grains. The precipitates in the diffusion zone are chromium-rich, surrounded by area of high concentration of niobium and molybdenum.

5. Conclusions

The process of explosive welding using ammonium nitrate (V) as the explosive, with detonation speed of ca. 2700m/s enabled obtaining joint between steel P355NH and Inconel 625 alloy. The joint was found to include precipitate zones with an increased concentration of imperfections in the form of cracks, voids and fragments of the surface layer of steel P355NH. Stress relief annealing (620°C / 90 min) led to partial recrystallization of steel P355NH in the joint area. Heat treatment in the form of normalizing (910°C/ 30 min) resulted in complete recrystallization of both welded materials, which was additionally confirmed by distributions of microhardness. Moreover, a diffusion zone of ca. 20 μm thickness formed in the joint and precipitates in the form of carbides of M_6C and $M_{23}C_6$ types appeared in Inconel 625 alloy. The presence of the chromium-rich precipitates have been stated in the diffusion zone.

6. Acknowledgements

This work uses results of the research from the project M-Era.net "Novel explosive welded corrosion resistant clad materials for geothermal plants", co-financed from The National Centre for Research and Development of Poland (No DZP/M-ERA.NET-2013/2309/2014). This work was also supported by the university grant RMN 725/2017.

References

1. S.N. Karlsdóttir, S.M. Hjaltason, K.R. Ragnarsdóttir, Geothermics **70**, 222-229 (2017)
2. G. V. Tomarov, A. A. Shipkov, Thermal Engineering **53**, 188-194 (2006)
3. K. Yang-Hyun, Y. Jae Ho, P. Jeong-Yong, K. Keon-Sik, K. Hyun-Gil, K. Dong-Joo, J. Yang-Il, S. Kun-Woo, Nuclear technology **186**, 295-304 (2014)
4. A. Strasser, J. Santucci, K. Lindquist, W. Yario, G. Stern, L. Goldstein, L. Joseph, *Evaluation of stainless steel cladding for use in current design LWRs. Final report* (Electric Power Research Institute EPRI NP, 1982)
5. W. Bristowe, M. Pearson, C. Stunguris, S. Gothard, *Titanium 2010 - Proceedings of the 26th Annual Conference of the International Titanium Association* (2010)
6. EN 10216-3:2002: *Seamless steel tubes for pressure purposes. Technical delivery conditions. Alloy fine grain steel tubes.*
7. J. C. M. Farrer, *The Alloy Tree: A Guide to Low-Alloy Steels, Stainless Steels, and Nickel-base Alloys* (Woodhead Publishing, 2004)
8. H. R. Z. Rajani, S. A. A. A. Mousavia, Combustion, Explosion, and Shock Waves **49**, 2 (2013)
9. H. R. Z. Rajani, S. A. A. A. Mousavia, Journal of Failure Analysis and Prevention **12**, 6 (2012)
10. H. R. Z. Rajani, S. A. A. A. Mousavia, Materials Science and Engineering **556** (2012).
11. H. R. Z. Rajani, S. A. A. A. Mousavia, Materials and Design **43** (2013)

12. W. Walczak, *Zgrzewanie wybuchowe metali*, (Wydawnictwa Naukowo-Techniczne, Warsaw, Poland, 1989)
13. T. Z. Blazynski, Applied Science (1983)
14. W. Babul, S. Ziemba, *Materiały wybuchowe w technologicznych procesach obróbki tworzyw*, (Państwowe Wydawnictwo Naukowe, Warsaw, Poland, 1972)
15. W. Babul, *Odkształcanie metali wybuchem*, (Wydawnictwa Naukowo-Techniczne, Warsaw, Poland, 1980)
16. H. Dyja, A. Maranda, R. Trębiński, *Zastosowania technologii wybuchowych w inżynierii materiałowej*, (Wydawnictwo Wydziału Metalurgii i Inżynierii Materiałowej Politechniki Częstochowskiej, Częstochowa, Poland, 2001)
17. T. Babul, P. Oleszczak, Inżynieria materiałowa **31**, 4 (2010)
18. F. Findik, Materials and Design **32** (2011)
19. A. Ambroziak, *Zgrzewanie tarciowe materiałów o różnych właściwościach*, (Oficyna Wydawnicza Politechniki Wrocławskiej, Wrocław, Poland 2011)
20. K. Topolski, P. Wieciński, Z. Szulc, A. Gałka, H. Grabacz, Materials and Design **63** (2014)
21. T. N. Prasanthi, C. Sudha, R. Kirana, S. Saibaba, Materials and Design **93**, 180–193 (2016)
22. R. Kosturek, M. Najwer, P. Nieslony, M. Wachowski, Advances in Manufacturing, Lecture Notes in Mechanical Engineering book series, 681-686 (2018)
23. A. Pocica, R. Bański, P. Waindok, Z. Szulc, A. Gałka *XVI International Conference ,Spawanie w energetyce", Poland* (2008)
24. H. Jian, X. Yan, J. Liu, X. Duan, Transactions of Nonferrous Metals Society of China **24**, 697-704 (2014)
25. S.A.A. Akbari Mousavi, P. Farhadi Sartangi, Materials Science and Engineering: A **494**, 329-336 (2008)
26. L. F. Trueb, Metallurgical Transactions **2**, 145-153, (1971)
27. J. Maliutina, V. Mali, A. K. Skorokhod, A. Bataev, Applied Mechanics and Materials **698,** 495-500 (2015)
28. F. Findik, R. Yilmaz, T. Somyurek, Scientific Research and Essays **6**, 4141-4151 (2011)
29. D.M. Fronczek, R. Chulist, L. Litynska-Dobrzynska, S. Kac, N. Schell, Z. Kania, Z. Szulc,J. Wojewoda-Budka, Materials & Design **130**, 120-130 (2017)
30. D.M. Fronczek, R. Chulist, Z. Szulc, J. Wojewoda-Budka, Materials Letters **198**, 160-163 (2017)
31. X. Wang, X. Li, F. Yan, C. Wang, Welding in the World **61**, 187-196 (2017)
32. C. Chen, H. Chen, W. Hwang, Materials Transactions **47**, 1232-1239 (2006)
33. M. Mazar Atabaki, M. Nikodinovski, P. Chenier, J. Ma, M. Harooni, R. Kovacevic, Journal for Manufacturing Science and Production **14**, 59–78.
34. P. Petrzak, K. Kowalski, M. Blicharski, Acta Physica Polonica A **130**, 4 (2016)
35. L.E. Shoemaker, The Minerals, Metals & Materials Society (2005)
36. A. Sukumaran, R.K. Gupta, V. Anil Kumar, Journal of Materials Engineering and Performance (2017)
37. P. Maj, B. Adamczyk-Cieslak, M. Slesik, J. Mizera, T. Pieja, J. Sieniawski, T. Gancarczyk, S. Dudek, Archives of Metallurgy and Materials **62**, 1695-1702.
38. L. Rongbin, Y. Mei, L. Wenchang, H. Xianchang, Journal of Materials Engineering and Performance, Volume **11** (2002)
39. W. C. Liu, F. R. Xiao, M. Yao, H. Yuan, Z. L. Chen, Z. Q. Jiang, S. G. Wang, W. H. Li, Journal of Materials Science Letters **17**, 245-247 (1998)
40. M. Yunpeng, L. Yongchang, L. Chenxi, L. Chong, Y. Liming, G. Qianying, L. Huijun, Journal of Alloys and Compounds **649**, 949-960 (2015)
41. B. López, I. Gutiérrez, J.J. Urcola, Diffusion Bonding **2**, 37-48 (1991)
42. J. L. Swartzendruber, Journal of Phase Equilibria **12** (1991)

An effective methodology to design scale model of non-metallic structural entity

Liming Yuan [a], Zhijie Xie, Fei Dai, Yonggang Xu, Yuan Zhang

Science and Technology on Electromagnetic Scattering Laboratory, Shanghai 200438, China

Abstract. In order to solve the problem on designing scale model of non-metallic structural entity, a method is first proposed according to the high-frequency approximation algorithm, in which reflection coefficients of designed scale model are optimized to be as identical as possible with those of theoretical scale model. An example is given to verify the effectiveness of the method, where a non-metallic cone is constructed and FEKO is employed to simulate the monostatic radar Cross Section (RCS) of the designed scale cone and the theoretical scale cone. Result reveals that the monostatic RCS of the designed scale cone agrees very well with that of the theoretical scale cone. The mean deviation of the monostatic RCS is just 1.31 dB in simulated elevation angles of 0°~90°. Thus, the method proposed in this paper is of important theoretical and practical significance in constructing scale model of high accuracy.

1 Introduction

Scale measurement technology has very important significance in studying electromagnetic scattering properties [1, 2]. In scale measurement, one of keys is to construct an accurate scale model. Existing researches begin with studying scale models of perfect electric conductor (PEC) and then come to study scale models of non-metallic materials as more and more composites are applied to objects [3-7]. However, the non-metallic materials in the existing studies have plated-shape without special circumstances. The non-metallic materials are always mounted on object surface and the scale materials can be designed according to the reflective loss. As the development of manufacturing technologies of composites, more and more non-metallic structural entities could be found in objects such as vehicles, aircraft, ships and so on. These non-metallic entities have great influences on the scattering properties. Meanwhile, these non-metallic entities not only have complicated electromagnetic properties but also have special geometric shapes. As it is always very difficult to manufacture scale material which has the same electromagnetic parameters with theoretical scale material, there is hardly any solution to solve the problem on constructing scale model of non-metallic structural entity over years. In this paper, a method is first proposed to solve the problem according to the high-frequency approximation algorithm. In order to verify the effectiveness of the designed result by the method, a non-metallic cone is constructed and FEKO is employed to simulate the monostatic radar Cross Section (RCS) of the designed scale cone and the theoretical scale cone.

2 Theories and methods

According to the high-frequency approximation algorithm, the scattering field can be calculated by the Stratton-Chu formula [8], which is expressed as follows,

$$\mathbf{E}^s = \frac{jk}{4\pi}\frac{e^{-jkr}}{r}\int_s \hat{s}\times(\mathbf{M}_s + \eta_0\hat{s}\times\mathbf{J}_s)e^{jk\mathbf{r}'\cdot(\hat{s}-\hat{i})}ds' \quad (1a)$$

$$\mathbf{H}^s = -\frac{jk}{4\pi}\frac{e^{-jkr}}{r}\int_s \hat{s}\times(\mathbf{J}_s - \frac{1}{\eta_0}\hat{s}\times\mathbf{M}_s)e^{jk\mathbf{r}'\cdot(\hat{s}-\hat{i})}ds' \quad (1b)$$

where $\mathbf{E}^s$ and $\mathbf{H}^s$ are scattering electric and magnetic field, respectively; $\mathbf{M}_s$ and $\mathbf{J}_s$ are induced electric and magnetic current on the surface, respectively; η_0 is the impedance in free space; k is the propagation constant in free space; r is the distance; $\hat{s}$ and $\hat{i}$ are unit vectors in scattering direction and incident direction, respectively; S is the entire surface of target. The induced electric current and magnetic current are calculated by the following expressions,

$$\mathbf{J}_s = \left[-(1-\Gamma_\perp)(\hat{\mathbf{e}}_\perp\cdot\hat{\mathbf{e}}_\mathrm{i})(\hat{\mathbf{i}}\cdot\hat{\mathbf{n}})\hat{\mathbf{e}}_\perp E^i + (1+\Gamma_\parallel)(\hat{\mathbf{e}}_\parallel\cdot\hat{\mathbf{e}}_\mathrm{i})\hat{\mathbf{n}}\times\hat{\mathbf{e}}_\perp E^i\right]/\eta_0 \quad (2a)$$

$$\mathbf{M}_s = (1+\Gamma_\perp)(\hat{\mathbf{e}}_\perp\cdot\hat{\mathbf{e}}_\mathrm{i})\hat{\mathbf{n}}\times\hat{\mathbf{e}}_\perp E^i + (1-\Gamma_\parallel)(\hat{\mathbf{e}}_\parallel\cdot\hat{\mathbf{e}}_\mathrm{i})(\hat{\mathbf{n}}\cdot\hat{\mathbf{i}})\hat{\mathbf{e}}_\perp E^i \quad (2b)$$

[a] Corresponding author: lming_y@163.com (Liming Yuan)

in which, $\hat{\mathbf{n}}$, $\hat{\mathbf{i}}$ and $\hat{\mathbf{s}}$ are unit vectors in normal direction of surface, incident direction and scattering direction, respectively; $\hat{\mathbf{e}}_i$ is the unit vector of incident electric field; $\hat{\mathbf{e}}_{\perp}$ and $\hat{\mathbf{e}}_{\parallel}$ are unit vectors of incident electric field in incident plane and perpendicular to the incident plane; $\Gamma_{\perp}$ and $\Gamma_{\parallel}$ denote reflection coefficients irradiated by TM and TE wave, respectively. It can be seen that the same reflection coefficients lead to the same induced electric and magnetic current and then the same scattering property. When scale material is designed by reflection coefficients, both $\Gamma_{\perp}$ and $\Gamma_{\parallel}$ should be taken into account. They can be calculated by following expressions,

$$\Gamma_{\perp} = \frac{\cos\theta_i - \sqrt{\varepsilon_r\mu_r - \sin^2\theta_i}}{\cos\theta_i + \sqrt{\varepsilon_r\mu_r - \sin^2\theta_i}} \quad (3a)$$

$$\Gamma_{\parallel} = \frac{\varepsilon_r\mu_r\cos\theta_i - \sqrt{\varepsilon_r\mu_r - \sin^2\theta_i}}{\varepsilon_r\mu_r\cos\theta_i + \sqrt{\varepsilon_r\mu_r - \sin^2\theta_i}} \quad (3b)$$

in which, θ_i is incident angle; ε_r and μ_r are relative permittivity and permeability of material. According to Eq. (3a) and (3b), scale model of non-metallic entity can be designed by following four steps.

(i) Input reflection coefficients of theoretical scale material. The reflection coefficients are calculated at different incident angles.

(ii) Built a collection of electromagnetic parameters at scale frequency. Selected some inclusions and mix them with a binder in different proportions. Measure electromagnetic parameters of mixtures and fit the measured data by effective medium theories. Built the collection of electromagnetic parameters by the fitting formulae and number each pair of permittivity and permeability.

(iii) Perform loop calculations. In each loop calculation, permittivity and permittivity are determined by a number generated by a random function. Calculate reflection coefficients and deviation by comparing with the reflection coefficients of theoretical scale material. Record the deviation and the corresponding electromagnetic parameters.

(iv) Search the optimal result. After the loop calculations, find out the optimal result by the minimum deviation of calculated reflection coefficients. The scale material could be determined according to the record in step (ii) and the designed scale model has the same geometric parameters with the theoretical scale model.

3 Results and discussions

Now an example is given to verify the effectiveness of the method. It is hypothesized that a non-metallic material is non-magnetic and its permittivity is 1-100j. The measurement frequency in full size system is 10 GHz and then the corresponding frequency in scale system is 100 GHz as the scale ratio is 1/10. The scale material can be designed by the method introduced in section 2. Here, the collection of electromagnetic parameters at 100 GHz is built by random functions. The collection includes 1000 pairs of permittivity ($\varepsilon_r = \varepsilon_r' - j\varepsilon_r''$) and permeability ($\mu_r = \mu_r' - j\mu_r''$). All ε_r', ε_r'', μ_r' and μ_r'' are in 1~50, 0~5, 1~5 and 1~2, respectively. Record and number each pair of permittivity and permeability. After optimizing, the permittivity and permeability of designed scale material are 47.66-1.86j and 0.29-1.89j, respectively. Figure 1 plots the complex reflection coefficients of the designed scale material and the theoretical scale material irradiated by TM wave and TE wave as the incidence angle changes from 0° to 90°. As shown in Fig. 1, it can be seen that the reflection coefficients of the designed scale material are very identical with those of the theoretical scale material.

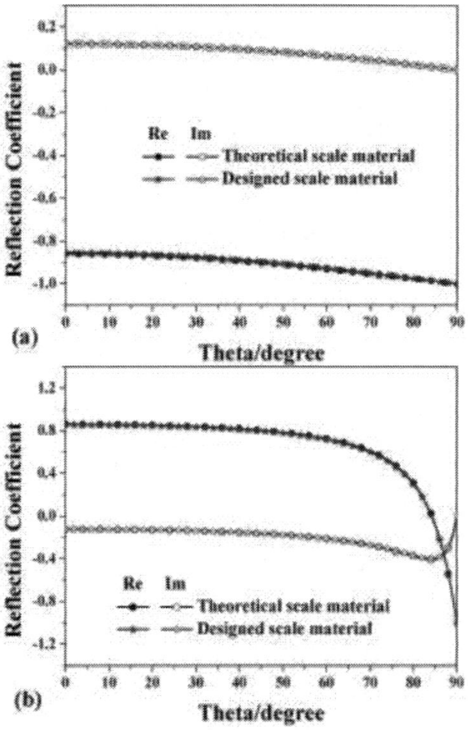

Fig. 1 The complex reflection coefficients of the theoretical scale material and the designed scale material irradiated by TM wave (a) and TE wave (b).

As introduced in section 2, designed scale model of non-metallic structural entity has the same geometric parameters with theoretical scale model. Next, a non-metallic cone is constructed and its monostatic RCS is simulated by commercial software FEKO. The non-metallic cone has a height of 500 mm and a PEC bottom of a 1000 mm diameter. The cone is comprised of non-magnetic material and the permittivity except the bottom is 1-100j. The designed scale cone and the theoretical scale cone have the same geometric parameters which are 1/10 of the corresponding parameters of the full size cone. The electromagnetic parameters of the theoretical scale cone are the same with those of the full size cone and the designed scale cone has the permittivity of 47.66-1.86j and the permeability of 0.29-1.89j. In simulations, monostatic RCS is calculated; TE wave is selected; the

[a] Corresponding author: lming_y@163.com (Liming Yuan)

frequency is 100 GHz; the azimuth angle is 0° and the elevation angle changes from 0° to 90°.

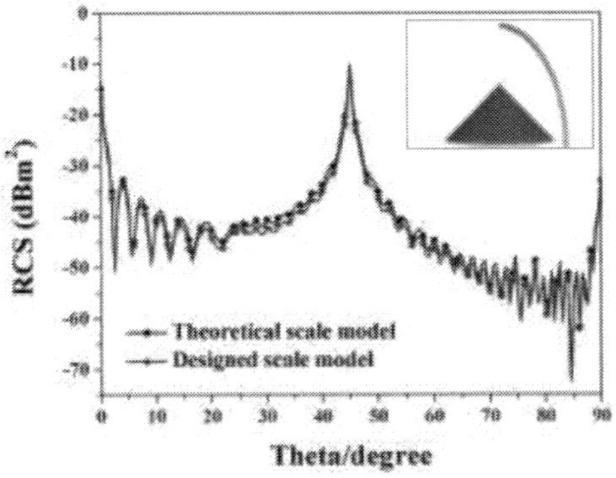

Fig. 2 The simulated RCS of the designed scale model and the theoretical scale model of the non-metallic cone.

Figure 2 shows the simulated RCS of the designed scale model and the theoretical scale model of the non-metallic cone. As shown in the figure, it can be seen that there is a good agreement on monostatic RCS between the designed scale cone and the theoretical scale cone. The average value of absolute deviations is just 1.31 dB in 0°~90°. And it is 1.20 dB, 1.42 dB and 1.31 dB in 0°~30°, 30°~60° and 60°~90°, respectively. The deviation mainly comes from non-plane scattering sources but it is completely in the range of RCS measurement tolerance.

4 Conclusions

According to the high-frequency approximation algorithm, a method is first proposed to design scale model of non-metallic structural entity. In order to verify the effectiveness of the method, an example is given to design scale model of a non-metallic cone. The simulated RCS of the designed scale cone is compared with that of the theoretical scale cone and the average value of absolute deviations is just 1.31 dB in 0°~90°. The deviation completely falls within the RCS measurement tolerance. Therefore, it can be concluded that the method proposed in this paper is very effective to design scale model of non-metallic structural entity.

Acknowledgements

This work was supported by Shanghai Municipal Science and Technology Commission under Grant Nos.18ZR1437400, 17YF1418400, 17QB1404100, 17210730900 and 16XD1423300, and the National Civil Aircraft Project of China under Grant MJZ-2017-F-11.

References

[1] White M O 1998 *Electronics & communication engineering journal* **10** 169
[2] Zhao Y et al 2012 *IEEE Transactions on Antennas and Propagation* **60** 5890
[3] Bird D 1994 *Radar Conference, 1994., Record of the 1994 IEEE National*, March 29-31, 1994, Atlanta, GA, USA, p. 74
[4] De Adana F S et al 2003 *IEE Proceedings-Radar, Sonar and Navigation* **150** 375
[5] Susetio A et al 2015 *Antennas and Propagation (APCAP), 2015 IEEE 4th Asia-Pacific Conference*, June 30 - July 3, 2015, Kuta, Indonesia, p. 568
[6] Yuan L M et al 2018 *Chinese Physics B* **27** 44101
[7] Xu Y G et al 2018 *J. Phys. D: Appl. Phys.* **51** 065004
[8] Lee J S et al 2013 *Antennas and Propagation Society International Symposium (APSURSI) 2013 IEEE*, July 7-13, 2013, Orlando, FL, USA, p. 606

The spheroidizing tendency and spheroidization mechanism of 75Cr1 steel

Jiping Huo, Chong Zhou and Hua Wang[a*]

Instrumental Analysis and Research Center of Shanghai University, Shanghai, China

Abstract. Spheroidization annealing process was carried out on 75Cr1 steel after cold rolling. Using TEM, SEM and image analysis software Image-Pro Plus 6.0, the effects of cold rolling and spheroidization processes on microstructure and mechanical properties were analyzed. The result shows that the cold rolling leads to fragmentation of the cementite and formation of the defects, which induces faster lamellae break-up and accelerates the annealing process. With the increase of isothermal temperature and holding time, the dimension of the spheroidized carbides increase as driven by Ostwald ripening mechanism, whereas the hardness of the steel decreases. The prior cold rolling before annealing process causes the yield strength improve slightly, however shows little effect on the cold formability. The optimum heat treatment process is, holding the samples at 705°C for 6 hours with cold deformation ratio (ε) of 0.6.

1 Introduction

75Cr1 steel, which belongs to Germany 1.2003 grade, is a cold-work alloy tool steel by German standard and mainly used in the manufacture of small tools. The high value in hardenability and toughness, and good property in shearing and wearing resistance of 75Cr1 steel makes it preferential in tool steel using[1-2], however, the high hardness and strength of 75Cr1 steel deteriorate the plasticity, thus a spheroidization annealing is required prior to industrial production to improve plasticity [3-4].

Many researches were found investigated on the quenching and tempering heat treatment processes of 75Cr1 steel, however few result is reported about the spheroidization annealing process. As been reported, the deformation usually accelerate the rate of spheroidization annealing and increase the spheroidization degree[5-6]. Nowdays most of the manufactures provide cold-rolled steel plates to company so as to reducing the cost of subsequent production and achieving rapid completion of spheroidization. In this paper, the effect of prior cold rolling on annealing process in 75Cr1 steel was investigated, which was expected to provide a theoretical base for annealing process design in 75Cr1 steel.

2 Experimental procedures

The chemical composition of the 75Cr1 steel was given in Table 1. A_{c1} and A_{c3} were detected to be 730°C and 753°C. The microstructure was detected using TEM JEM-2100 and SEM JEM-6700. At first, the cold rolled sample with the intermediate deformation (ε=0.4) was used as the research object in determining the spheroidization annealing temperature and holding time. Then, the spheroidization annealing of different cold rolling deformation samples (ε=0, 0.2, 0.4, 0.6) were studied. The annealing process were simulated in a CARBOLITE tube furnace, the accuracy of the heat controller was ±5 °C. The heat treatment process was shown in Fig. 1.

Table 1. Chemical composition of the 75Cr1 steel (mass %)

C	Si	Mn	P	S	Cr	Cu	V	Ti
0.75	0.29	0.77	0.01	0.002	0.46	0.02	0.003	0.004

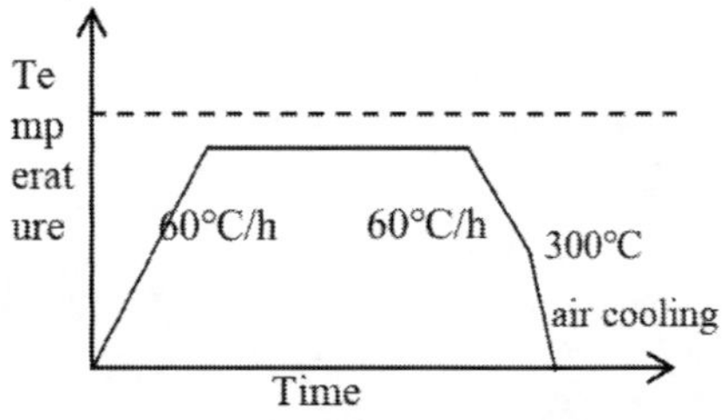

Figure 1. Spheroidizing annealing process of the investigated steels

Metallographic specimens were firstly prepared according to the standard procedure of the spheroidized sample and then etched with nital (2% HNO3 in ethanol) for a few seconds. The vickers hardness of all specimens were measured with a loading force of 0.5 kgF (4.905 N) with a dwell time of 10 s. The tensile test was conducted at room temperature according to ISO 6892:1998 standard, during which, the test speed was 30 mm/min.

[a] Corresponding author: wh225@shu.edu.cn

 https://doi.org/10.1051/matecconf/201824201009

The degree of spheroidization was quantified by image analysis over randomly selected SEM micrographs (at 7000 times magnification). Particles with an area smaller than 0.001μm² were not considered in the analysis. Cementite particles with an aspect ratio lower than 2 were considered to be spheroidized in this study. Aspect ratio is defined as the relation between the length and the width of the cementite, and it is provided by the image analysis software Image-Pro Plus6.0 [7-8].

3 Results and discussions

3.1 Raw materials

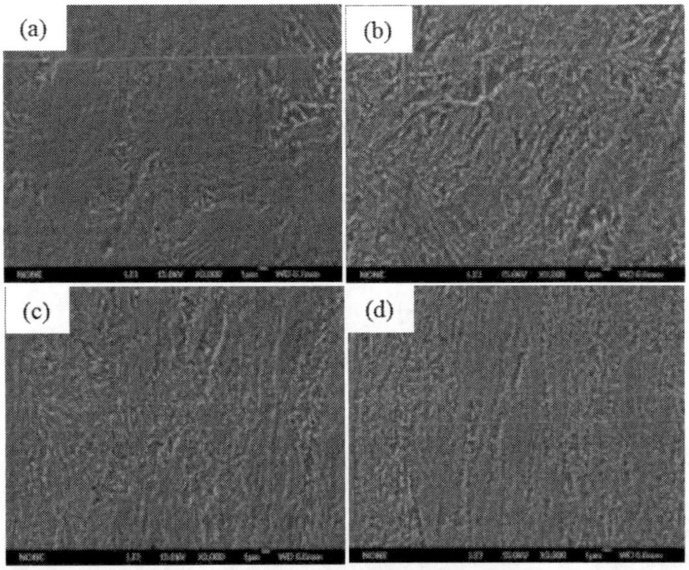

Figure 2. The microstructure of 75Cr1 steel with different cold rolling deformations(a) ε=0、(b) ε=0.2、(c) ε=0.4、(d) ε=0.6

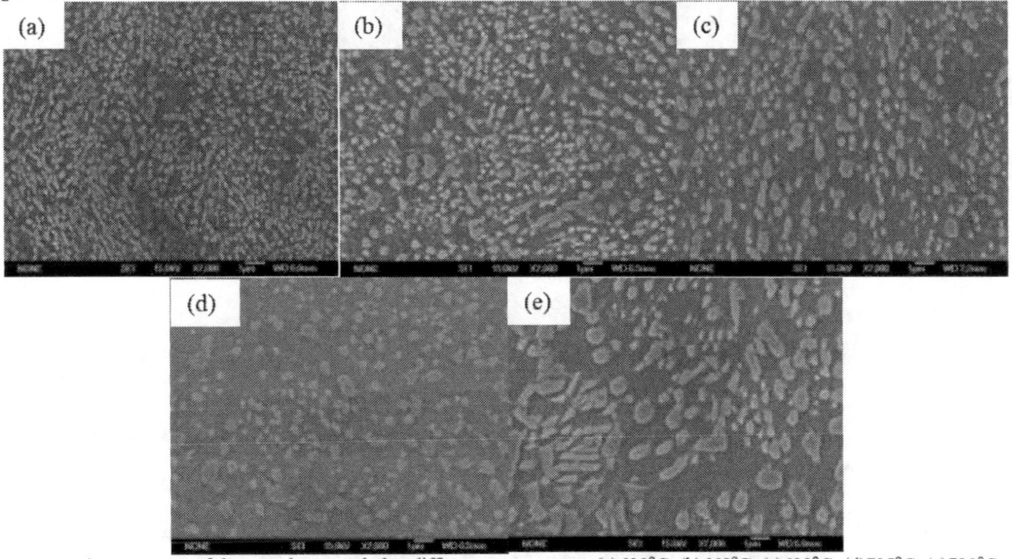

Figure 3. Microstructure of the samples annealed at different temperatures (a)630°C, (b)660°C, (c)690°C, (d)705°C, (e)720°C

With the cold rolling deformation increased, the hardness increased because of the deformation strengthening, as shown in Table 2. The SEM microstructure with different cold rolling deformations were shown in Fig. 2. Because the pearlites in the original structure had different orientations, the cold rolling made the microstructure transformed through a process of "uniform→uneven →uniform". In the case that pearlite interlaminar has a large intersection angle with the strain direction, the microstructure will be refined during cold deformation. On the contrary, if the pearlite interlaminar parallels to the strain direction, the cementite would be bent and fractured, as shown in Fig. 2c. It can be seen from Fig. 2d nearly all the pearlite structure were parallel to the strain direction, and the microstructure was refined as the deformation reached up to 60% [9].

Table 2. The hardness of 75Cr1 steel under different cold rolling deformations

Deformation%	0	0.2	0.4	0.6
Hardness HV	298.0	398.6	419.0	427.8

3.2 Effect of isothermal temperature on microstructure of the annealed steel

Isothermal temperature is one of the most important factors affects the spherdidization annealing process. Supposing the carbides volume fraction is same, the larger the carbides size is, the lower the steel hardness becomes[10]. In the present work, 75Cr1 steel samples (ε=0.4) were heated individually to 630oC, 660oC, 690oC, 705oC and 720oC, then preserved by a time of 6h. After heat preservation, the samples were cooled to 300oC at a rate of 1oC/min and then air-cooled to room temperature. The final microstructure was shown in Fig. 3.

Image-pro Plus6.0 was used to calculate diameter, aspect ratio and spheroidization rate of the SEM micrographs (at 7000 times magnification) at different isothermal temperature, and the results were shown in Figs. 4 and 5. The cementite particles showed an increasing tendency with the higher isothermal temperature, which was considered because the lamellae spheroidiztion kinetic was temperature dependent[11-13]. When the sample was annealed at 630°C, most of cementite particles were smaller than 0.2μm, however the globular shape of these particles with an aspect ratio more than 2. With the temperature continued to increase, the cementite particles with diameter less than 0.3μm decreases rapidly in number. After this sharp decrease, the production of globular particles slowed down and the globular particles became coarsen. When the sample was annealed at 705°C, the spheroidization rate grew up to 93.8%, however, the spheroidization rate decreased to 89% at 720°C. It was indicated that the temperature improvement accelerated the diffusion of carbon and thus induced the spheroidization become faster, that was, made the small particles grew into larger particles by Ostwald mechanism. However, the temperature which was close to A_{c1}, will lead to the coarse of the particle and even induce some particle grew to lamellae, causing the spheroidization rate decreased. The hardness curve with respect to isothermal temperature after spheroidization annealing was shown in Fig. 6. With isothermal temperature increased, the hard phase number decreased and the hardness reduced. When the sample was annealed at 720°C, the minimum hardness was 191.2 HV, however the spheroidizing effect was terrible and the cementite coarsen heavily. Therefore, the reasonable spheroidization annealing temperature is 705°C.

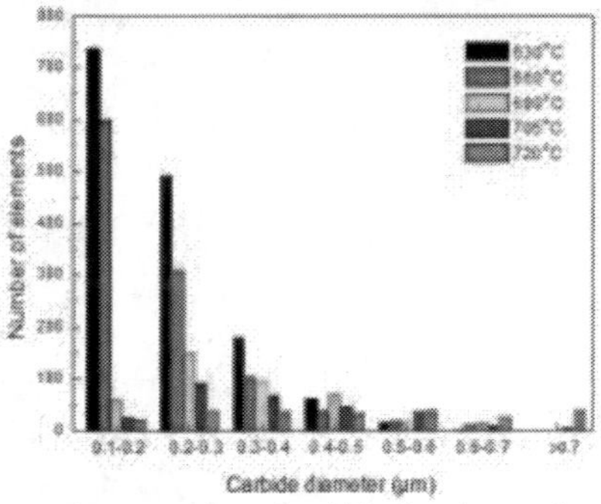

Figure 4. The arrange law of carbide sizes in samples with different annealing temperature

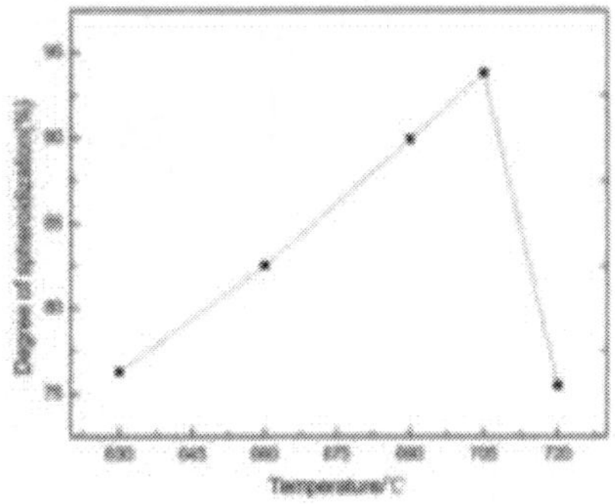

Figure 5. Evolution of the spheroidization degree in samples with respect to annealing temperature

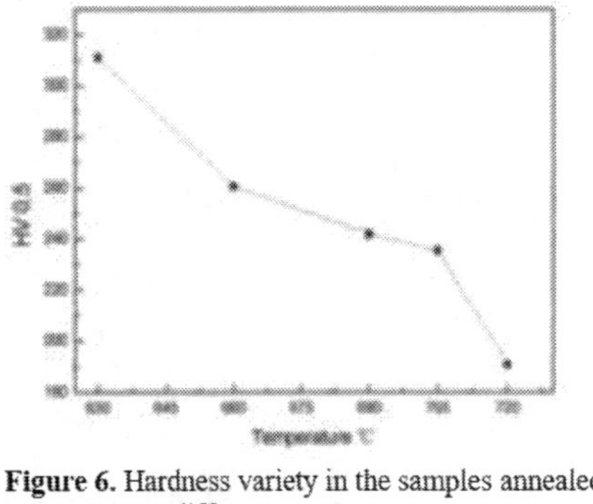

Figure 6. Hardness variety in the samples annealed at different temperature

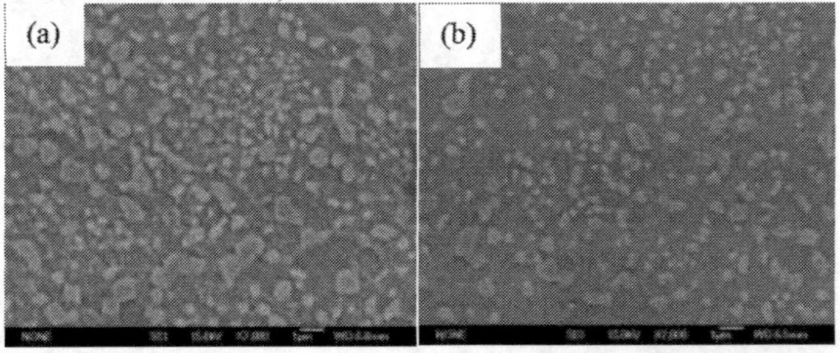

 https://doi.org/10.1051/matecconf/201824201009

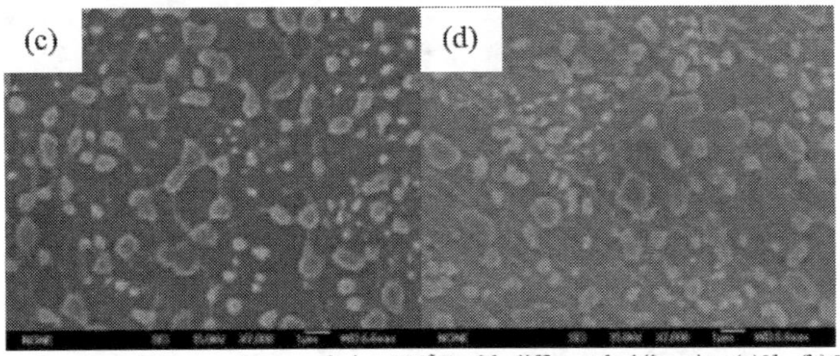

Figure 7. Microstructure of the sample annealed at 705°C with different holding time(a)3h, (b)6h,(c)9h, (d)12

3.3 Effect of spheroidizing time on the microstructure of annealed steel

During heat treatment, cementite lamellae was splited to globular units and coarsen as driven by the Ostwald ripening process, as shown in Fig. 7. It can be seen from Figs. 8 and 9 that the size of carbides in these samples increased with the extension of holding time. At the time of 3h, most of the particles were at the range of 0.1-0.2μm. The spheroidization degree was higher with time prolonged, but the spheroidization degree decreased a little when holding time was 12h due to the corasen of cementite.

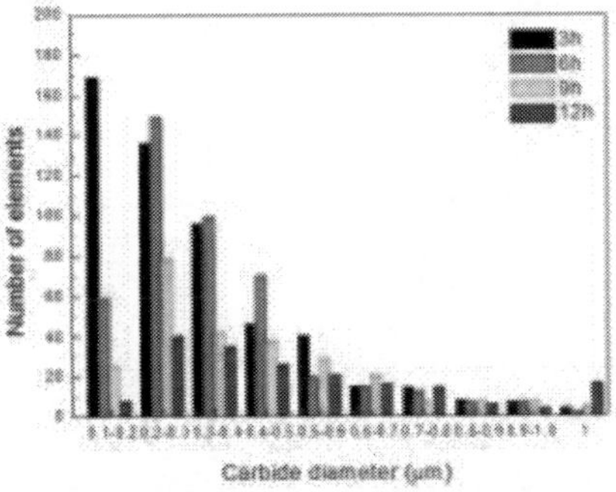

Figure 8. The arrange law of carbide sizes in samples with different holding temperature

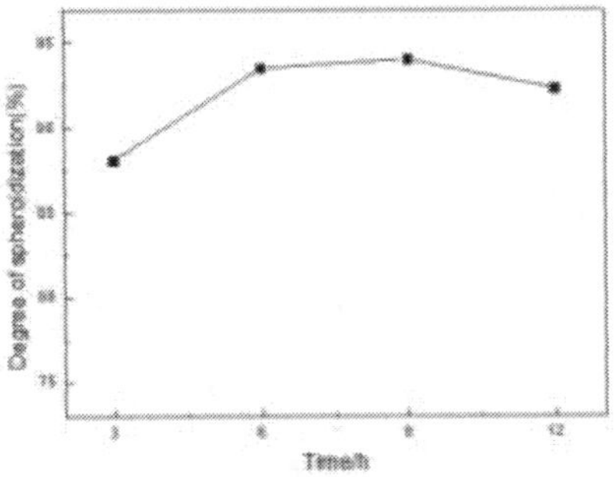

Figure9. Evolution of the spheroidization degree with different holding time at 705°C

With the holding time prolonging, some small carbides were dissolved, while the remained carbides trended to become the nucleation core of newly generated carbides. When holding time was short,the amount of small carbides (<0.3μm) was large because the lamellea carbides were dissolved into particles, and the hardness was higher than the other samples, as shown in Fig. 10. Although the hardness decreased when extending the holding time, it must be considered as well that abnormal grains may generated, which are harmful to cold formability[14]. Thus, the duration of the annealing should be adjusted with the aim of satisfying not only a minimum hardness but also a appropriate microstructure. When the holding time was 6h, the spheroidization rate was higher, the hardness reached the value (<250 HV) required for cold working, and the carbide size was uniform. Therefore, to get a proper microstructure, the holding time of 6h can be chosed for this heat treatmen.

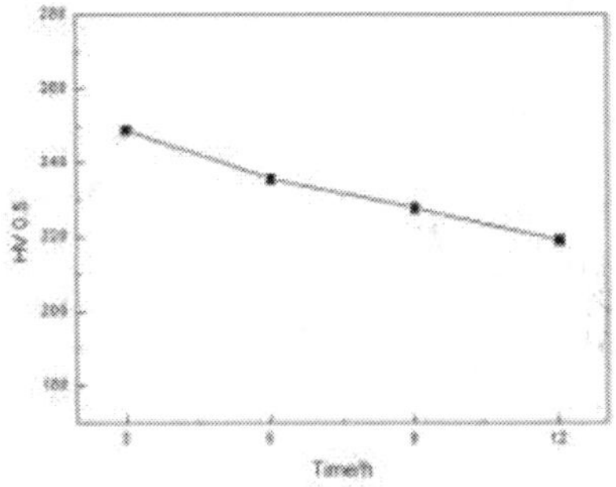

Figure10. Hardness variety in the samples annealed with different holding time at 705°C

3.4 Effect of cold rolling on the microstructure of annealed steel

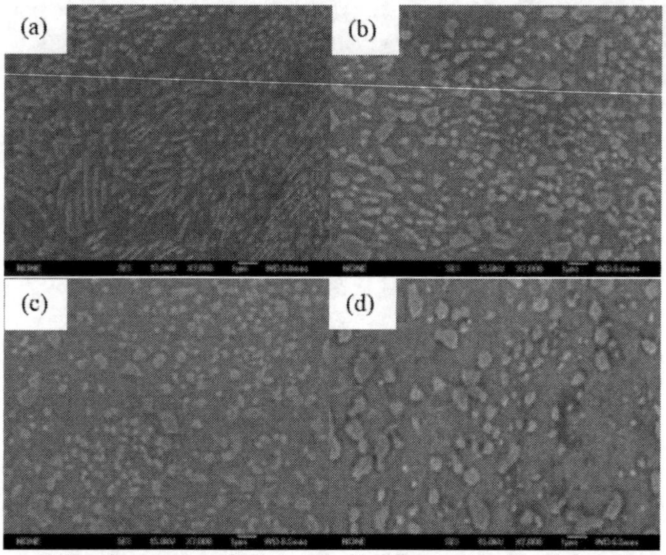

Figure 11. The microstructure of 75Cr1 steel with different cold rolling deformations that (a)ε=0, (b)ε=0.2, (c) ε=0.4, (d)ε=0.6 after annealing by 6 hours at 705°C

It is reported that besides defect migration and cementite break-up, cementite sub-boundary splitting also acclerates spheroidization[15-16]. The microstructure evolution after deformation at different ratio for 6h at 705°C was shown in Fig. 11. For the non-deformed sample, it was observed that several pearlite colonies still lied in the microstructures, despite the cementite lamellae being partially fragmented into short units, the spheroidization rate was the lowest in the four microstructures with different deformation ratio. It can be seen from Figs. 11 and 12, when the deformation continued to increase, the spheroidization became more complete. After spheroidizing, the cementite began to grow, and the diameter in samples with deformation ratio (ε) of 0.6 was significantly larger than that of 0.4, as shown in Figs. 11c and 11d. It can be concluded that the cold deformation can significantly shorten the spheroidization annealing time and accelerate the spheroidization process

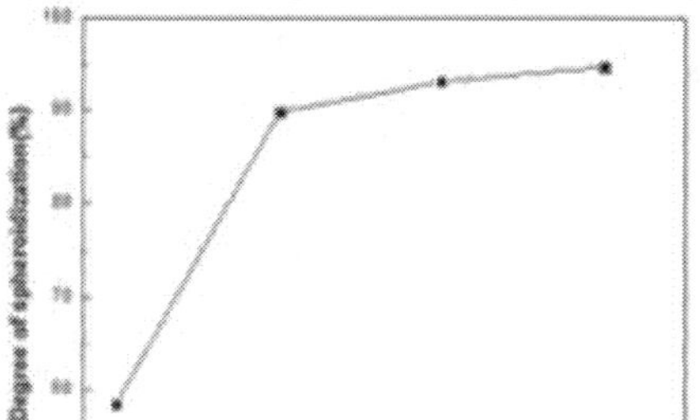

Figure12. Evolution of the spheroidization degree with different cold rolling deformations

3.5 The role of cold deformation in spheroidization annealing

During spheroidization annealing, the changes on the cementite are mainly through the mechanism of the defects migration in cementite lamella and the Rayleigh instability theory. The cementite in the spheroidization annealing process of the test steel after cold rolling mainly fragment through two mechanisms, one is that the cementite is broken by cold rolling process, and the other is that the cementite is broken due to the cementite dissolution during annealing[17-19]. As shown in Figs. 13b and 13f, the pearlite before deformation was laminae, and the diffraction patterns was regular. When ε was 0.6, the diffraction pattern of the cementite became circular, it was indicated that the cementite had rotated to different directions during the deformation process, and the interface between the ferrite and the cementite became blurred. The increase in cold rolling deformation caused a large number of defects inside the structure, generated deformation energy, and provided more channels for the diffusion of carbide, as shown in Fig. 13e.

It can be seen from Figs. 14, 15 and 16 that before deformation, the interlamellar pearlite space was 4.1nm and the thickness of cementite was about 0.6 nm. When ε was 0.2, part of pearlite had been broken into granules which became the nucleation core of newly generated carbides, but most of them still stayed as a laminated structure. When ε was 0.6, the cementite layer spacing was about 0.6 nm, and the thickness was reduced to 0.25 nm. According to the spheroidization kinetics, the relationship between the instantaneous growth rate of the carbide aggregation and the interlamellar pearlite spacing is shown in the equation (1) [20-21]:

$$\frac{\partial r}{\partial t}=\frac{4D\Omega C_0\sigma}{RTrl_0}\cdot\frac{(1/r_1-1/r_2)}{(1/r_1-1/\lambda)} \quad (1)$$

Where D represents diffusion coefficient, C_0 represents the concentration of alloying elements in the matrix when the position is at the radius with zero curvature, Ω is molar volume, σ is the interfacial energy, T is absolute temperature, R is diffusion

[a] Corresponding author: wh225@shu.edu.cn

radius, r_1 and r_2 is the radius of diffusion, l_0 is the length of cementite, λ is the interlamellar pearlite spacing. When the cementite is finer and the interlamellar spacing is smaller, the diffusion distance of the alloy elements becomes short, and this is beneficial for the aggregation of the cementite. Therefore, the microstructure after cold rolling deformation is more favorable for the spheroidization of the cementite.

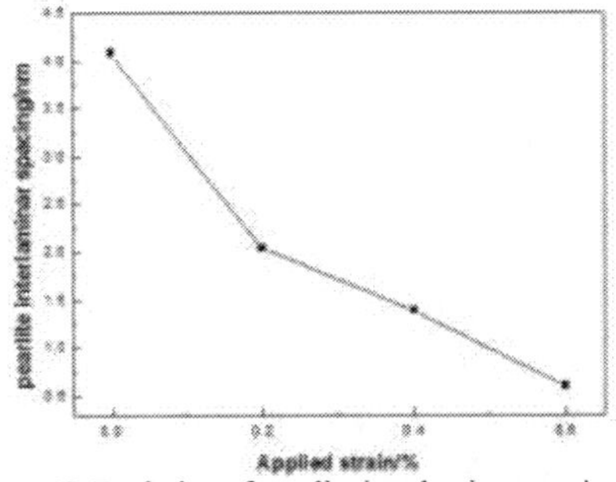

Figure 15. Evolution of pearlite interlaminar spacing with respect to cold rolling deformation

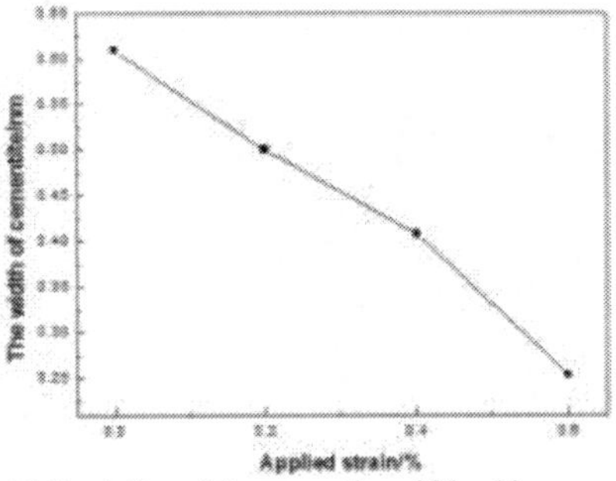

Figure16. Evolution of the cementite width with respect to cold rolling deformation

3.6 Mechanical prperities discussions

The mechanical properties of the experimental steel which were affected by grain size, cementite distribution and ferrite matrix were shown in Table 3. It can be seen that the tensile strength and yield strength increased at first, and then decreased to their minimum values of 223.3 HV and 680 MPa with the increase of deformation, and this tendency was same as that of hardness. However, the tensile elongation decreased first and then increased. The yield strength of samples with prior deformation was higher than non-deformed samples, it may be resulted from the presence of deformation energy. During these treatment, the deformation samples will recrystallize. When the cold rolling deformation increases, the grains become finer after recrystallization [22-23], some reseraches have reported that the grain size has weak effect on cold formability[24]. When ε was 0.2, the fine cementite particles formed by annealing were dispersed on the ferrite matrix which can contribute to the steel strength improvement through second phase strengthening, thus the tensile strength increased whereas the tensile decreased. Besides, the resistance to overcome when dislocation bypasses spherical carbides during stretching is the formula (2):

$$\tau = Gb/L \qquad (2)$$

Where G is the shear modulus of the material, b is the Burgundian vector, and L is the average space of the second phase. When L increases, τ decreases, and the value of tensile strength as well as yield strength reduces.

The tensile elongation of non-deformed sample was 11.78%. It can be seen from Fig. 17a that the distribution of dimple was uneven, even some area presented river pattern, thus non-deformed samples had unsatisfied performance in deformation ability. It was indicated that the deformation ability of the steel was also infuenced by the degree of spheroidization, and the cold formability cannot be considered only from the parameters such as hardness or cementite particle size [25]. When ε was 0.2, the dimple was the smallest and shallowest of the four SEM micrographs, and the cold formability was the worst of the four samples. The size of the dimple was equivalent to the carbide particles[26]. As the cold rolling deformation futher increased, the fracture dimple became larger, and the plasticity was better. When ε was 0.6, the hardness after annealing was 223.3 HV, the maximum tensile elongation was 13.01%, and the tensile strength was 680 MPa.

Table 3. Mechanical properties of steel sample after spheroidizing annealing with different cold rolling deformations

Applied strain	Tensile strength (MPa)	Yield strength (MPa)	Tensile Elongation (%)	HV
0	759	440	11.78	234.7
0.2	787	660	10.75	235.6
0.4	733	651	11.42	232.8
0.6	680	620	13.01	223.3

4 Conclusion

(1) The best spheroidization annealing process for 75Cr1 steel is as follows: Firstly, cold roll the steel with the deformation ratio (ε) of 0.6; then isothermal the steel at 705 °C for 6 hours. After this annealing process, the tensile strength of the steel is more than 680 MPa, the hardness is 223.3 HV, and the maximum elongation is 13.01%.

(2) During the spheroidization annealing process of the 75Cr1 steel, the carbide size increases in Ostwald mechanism, and the steel hardness is reduced from 311.4 HV to 191.2 HV. With the increase of isothermal temperature, the degree of spheroidization rate first increases rapidly to 93.8%, and then decreases to 76.3%.

(3) After cold rolling, not only a great many of small particle carbides in the 75Cr1 steel can directly contribute to spheroidization as the core of spheroidization annealing, but also a large number of dislocations and subgrain boundaries will acclerate the kinetics of spheroidization.

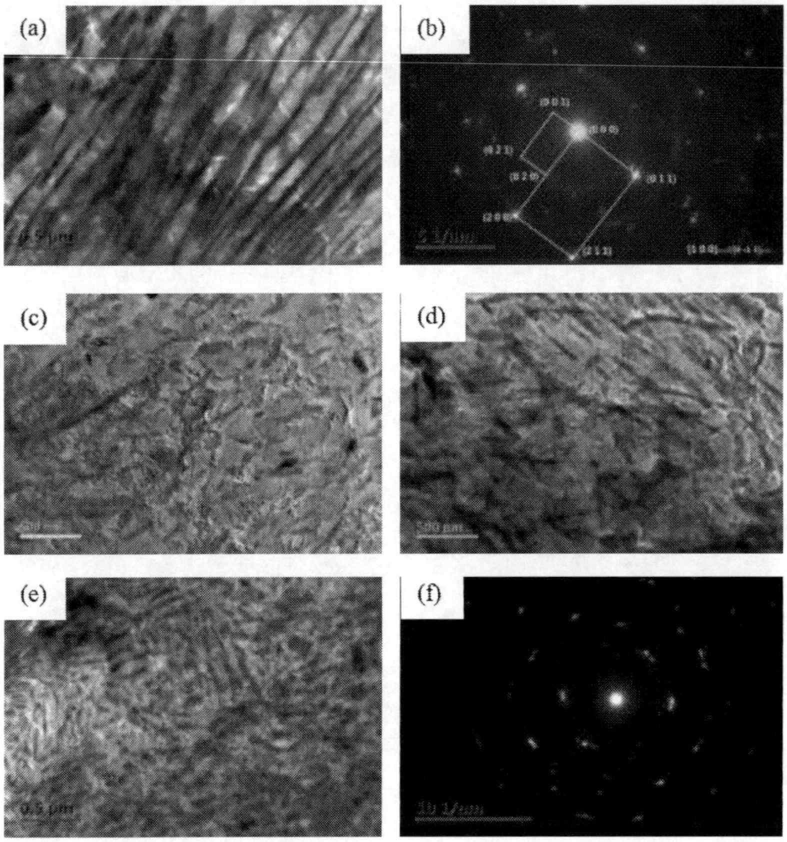

Figure13. The TEM microstructure and diffraction spot of 75Cr1 steel with different cold rolling deformations at high magnification;(a)(b)ε=0, (c) ε=0.2, (d) ε=0.4, (e)(f)ε=0.6

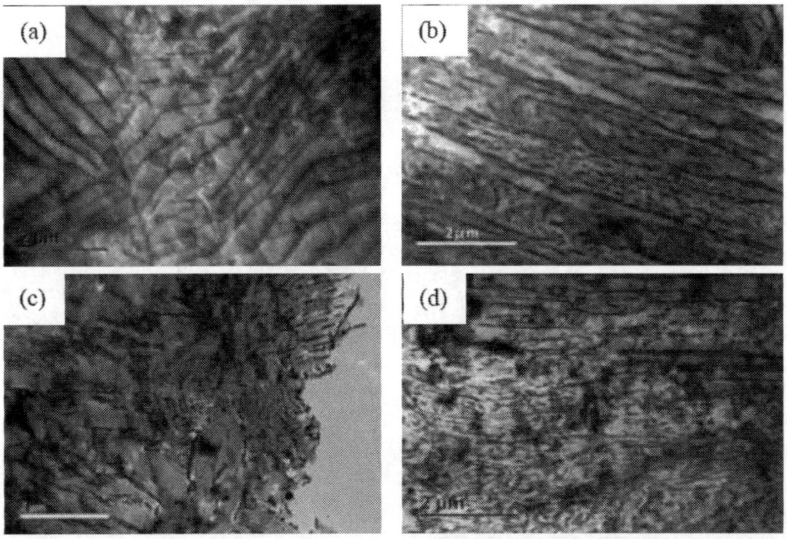

Figure14. The TEM microstructure of 75Cr1 steel with different cold rolling deformations at low magnification (a)ε=0, (b)ε=0.2, (c)ε=0.4, (d)ε=0.6

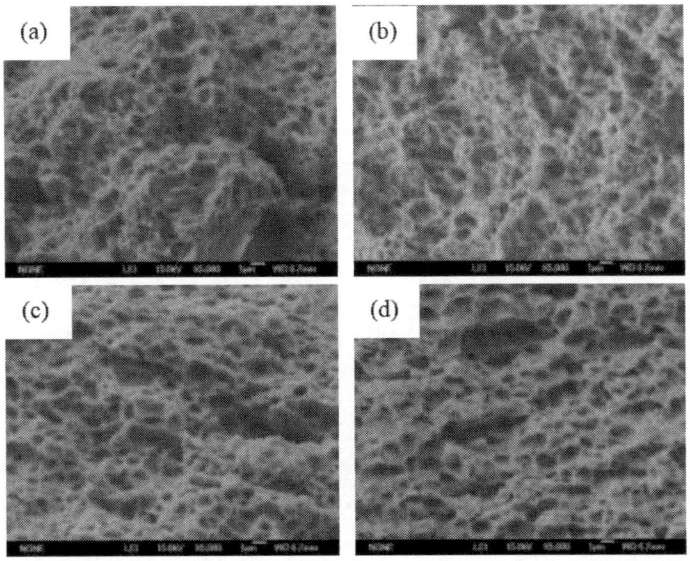

Figure 17. The SEM micrographs of tensile fracture of 75Cr1 steel with different cold rolling deformations that (a)ε=0, (b)ε=0.2, (c) ε=0.4, (d)ε=0.6 after annealing by 6 hours at 705°C

(4) After spheroidization annealing to 75Cr1 steel, the hardness decreases with the increase of caibide size, because the resistance will decrease when the dislocations bypassing the carbide.

(5) The yield strength of the cold rolled samples increases slightly compared with the un-deformed ones after annealing, because of the recrystallizing in them. The main factors that affect the cold formability are: spheroidization rate, cementite size and the distribution of cementite particles.

Reference

1. Tianjin Science and Technology Association, *Manual of Steel* (Beijing: Popular Science Press,1959)
2. X. T. Bo, Xing L. Heat Treatment Technology and Equipment **27** 14-7 (2006)
3. W. T. Yu, J. Li, C. B. Shi and Q. T. Zhu, Mater. Eng. Perform. **26** 1-10 (2016)
4. D. Yao, J. Li, J. H. Li and Q. T. Zhu, Mater. Manuf. Process. **30** 111–5 (2015)
5. M. Caruso, H. Verboomen and S. Godet, Advanced Materials Research **409** 666-71 (2012)
6. C. S. Zheng and L. F. Li, Heat Treatment of Metals **40** 31-34 (2015)
7. J. M. O'Brien and W. F. Hosford, Metall. Mater. Trans. A **33** 1255-61 (2002)
8. N. Nutal, C. J. Gommes, S. Blacher and P. Pouteau, Image. Anal. Stereol. **29** 91-8 (2011)
9. S. H. Sun, Y. Xiong, W. T. Fu, G. Z. Xing and Y. Z. Gu, Acta. Metall. Sin. **41** 267-70 (2005)
10. D. L. Li, D. S. Ma, Z. Z. Chen, H. Ji, B. S. Liu, A. J. Kuang and Z. L. Zhan, Heat Treat. Met. **35** 57–61 (2010)
11. Y. L. Tian, R. W. Kraft, Metall. Trans. A. **18** 1403-1414 (1998)
12. A. Lindsley and A. R. Marder, Acta Mater. **46** 341-51 (1997)
13. T. Sakuma, N. Watanabe and T. Nishizawa, Trans Jim **21** 159-68 (2007)
14. J. M. Rodriguez-Ibabe, Mater. Sci. Forum **284-286** 51-62 (1998)
15. C. L. Xing,Y. Peng, J. K. Xing and Y. Zang, Heat Treat. Met. **40** 94-98 (2015)
16. J. Arruabarrena, B. López and J. M. Rodriguez-Ibabe,Metall. Mater. Trans. A **45** 1470-84 (2014)
17. Y. T. Wang, Y. Adachi, K. Nakajima and Y. Sugimoto, Acta Mater. **58** 4849-58 (2010)
18. Y. L. Tian and R. W. Kraft, Metall. Trans. A **18** 1403-14 (1987)
19. G. Sharma, R. V. Ramanujan and G. P. Tiwari, Acta Mater. **48** 875-89 (2000)
20. H. Di, X. Zhang, G. Wang, X. J. Xiu, Mater. Process. Tech. **166** 359-63 (2005)
21. X. Wu, Z. X. Zhao, and R. D. Xue, Trans. Mater. Heat Treat. **10** 98–102 (2014)
22. Z. Q. Cui, B. X. Liu, *Metallography and Heat Treatment Theory* (Harbin Institute of Technology Press, 1998)
23. J. Arruabarrena, B. López and J. M. Rodriguez-Ibabe, Metall. Mater. Trans. A **45** 1470-84 (2014)
24. M. Calcagnotto, Y. Adachi, D. Ponge and D. Raabe, Acta Mater. **59** 658-70 (2011)
25. J. M. O'Brien and W. F. Hosford, Metall. Mater. Trans. A **33** 1255-61 (2002)
26. A. K. Vasudévan and R. D. Doherty, Acta Metall. **35** 1193-219 (1987)

Acknowledgements

The authors would like to acknowledge the support of the National Key R&D Program of China (No.2017YFB030 4402).

Prediction and verification of the single cladding layer thickness for laser solid forming

Jun-hua Wang [1,2], Fu-zhu Han [1,2,*] and Wei-sheng Ying [1,2]

[1] *Department of Mechanical Engineering, Tsinghua University, Beijing 100084, China*
[2] *Beijing Key Lab of Precision/Ultra-precision Manufacturing Equipments and Control, Tsinghua University, Beijing 100084, China*

Abstract. In order to accurately measure the powder effective utilization rate and to predict the single cladding layer thickness in laser solid forming process, a new method to measure powder distribution density is presented in this paper. Using this method, the powder material entering molten pool was collected by a copper tube with the same size as laser spot and its mass was measured to accurately calculate powder effective utilization rate in the laser solid forming process. And then, the single cladding layer thickness were calculated according to the conservation law of powder material mass before and after forming, and the correctness of calculating results was verified via a series of single track experiments. The results show that the theoretical calculation results of single cladding layer thickness are well coincident with the experimental results, and it proves that the measurement method of powder distribution density is accurate and reliable. In addition, laser power has a great influence on the geometry size of single cladding layer. Therefore, the influence of laser power on the geometry size of single cladding layer should be taken into account when establishing the prediction model of single cladding layer thickness.

Key words. Laser solid forming; Single cladding layer thickness; Powder effective utilization rate; Coaxial powder feeding; Powder concentration;

1 Introduction

Laser solid forming (LSF) is a new type of advanced manufacturing technology which can be used to rapidly fabricate the full dense complicated metallic components with high performance not using molds [1]. As to LSF technology, coaxial powder feeding technology is one of the most nuclear and key techniques [2]. In most cases, only a part of powder materials from powder feeder be sent to the molten pool and solidified into deposition materials because the size of powder flow is larger than that of laser spot. Therefore, it is necessary to explore powder effective utilization rate which is the ratio of the mass of powder materials entering the molten pool to that of powder materials from powder feeder per unit time [3]. Due to the mass of powder materials sent out by the powder feeder per unit time is known, whether the mass of powder materials entering the molten pool can be measured accurately has become the key to explore powder effective utilization rate.

Over the past decade, many researchers have studied powder flow distribution and its influences on the geometry size of single cladding layer. Typical method for powder flow distribution is based on optical methods. This method measure the intensity of the scattered light by the powder stream using a CCD sensor which provides the information to a digital image processing unit [4]. Due to the powder distribution has a positive correlation with the scattered light intensity, the powder distribution can be predicted according to the scattered light intensity. The main problem with optical methods is that they only obtain a qualitative concentration distribution [5]. This implies a great problem when the model is being validated, since it can only compare the powder flux distribution profiles without taking into account numerical values. Using this method, it is difficult to measure powder effective utilization rate accurately.

In order to accurately measure the powder effective utilization rate and to predict the single cladding layer thickness in laser solid forming process, this article proposes an alternative method to measure the mass of powder materials entering the molten pool. The methodology is relatively simple and is based on the weight measurement. Using this method, the powder material was collected by a copper tube with the same size as laser spot and its mass was measured to accurately calculate powder effective utilization rate. And then, the single cladding layer thickness were calculated according to the conservation law of powder material mass before and after forming, and the correctness of calculating results was verified via a series of single track experiments. The results show that the theoretical calculation results of single cladding layer thickness are well coincident with the experimental results.

* hanfuzhu@mail.tsinghua.edu.cn

2 Powder effective utilization rate

2.1 Powder distribution detection method

Figure.1 shows the schematic diagram of the density distribution of powder flow. In Figure.1, there are significant differences in the size of powder flow and its distribution density in the horizontal plane with different spacing to the powder nozzles. Different powder flow size and distribution density will result in significant difference of powder effective utilization rate.

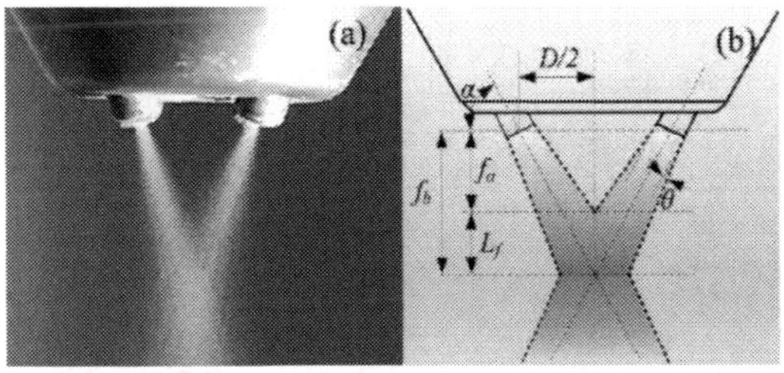

Figure 1. Schematic diagram of the density distribution of powder flow.

In order to accurately measure the powder effective utilization rate and to predict the single cladding layer thickness in LSF process, this article proposes a new detection method to measure the mass of powder materials entering the molten pool, its schematic diagram as shown in Figure.2(a). The methodology is relatively simple and is based on the weight measurement. Using this method, the powder material entering molten pool was collected by a copper tube with the same size as laser spot, and its mass was measured to accurately calculate the powder effective utilization rate. When the powder feeding system works, the powder particles falling into the test area freely fall into the powder container along the copper pipe. By measuring the mass of powder materials entering the container per unit time, the powder effective utilization rate in the test position can be calculated accurately. The distribution density of powder flow was determined by changing the test position in the horizontal plane with different spacing to the powder nozzles.

Using the method above, the distribution density in different horizontal planes were tested to clearly identify the distribution density of powder flow under coaxial powder feeding mode. The distance from the test horizontal planes to the nozzle was set to 11mm, 12mm, 14.5mm and 19mm, respectively. Measure specific time was set as 2min. In the experiments, the powder particles with the size of 75~185μm used in the experiments was martensitic stainless steel (17-4PH, density=7.8g/cm^3). The schematic diagram of measurement process and the measurement results of powder distribution density in X/Y axis directions of different horizontal planes are shown in Figure.2(b). As shown in Figure.2(b), there are significant differences in distribution density in X/Y axis directions of different horizontal planes. When H=14.5 mm (H=f_b), the test horizontal plane is the theoretical focal plane of coaxial powder feeding system. In this horizontal plane, the distribution density of powder flow in X/Y axis directions presents the Gaussian distribution feature, and the powder density reaches to the maximum. When H=11mm, the distribution density curves no longer obey the Gaussian distribution, but presents the double hump distribution characteristics.

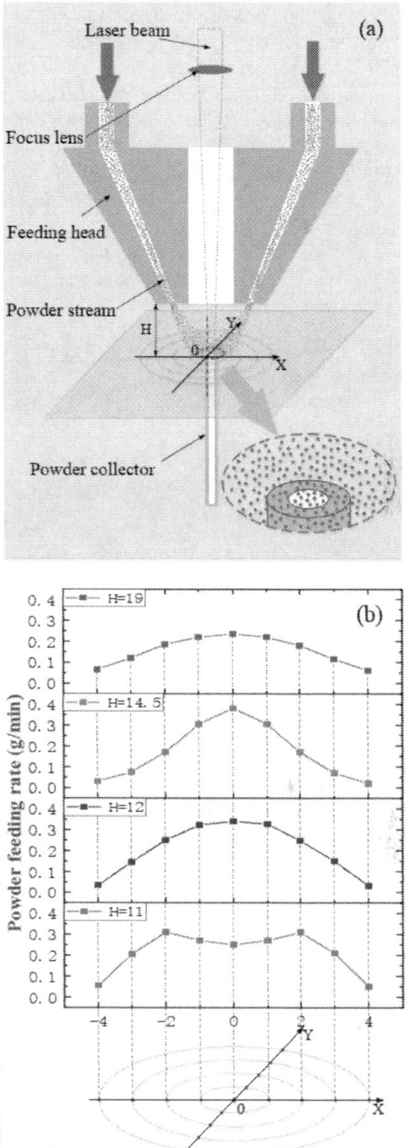

Figure 2. Schematic diagram of measurement process and the measurement results of powder distribution density in X/Y axis directions of different horizontal planes.

2.2 Measurements of powder effective utilization rate

According to the detection results of distribution density of powder flow in X/Y axis directions of different horizontal planes under the coaxial powder feeding mode, the powder distribution density varies significantly in the laser beam center in different horizontal planes. This will result in a significant difference in powder effective utilization rate.

The powder effective utilization rate varies in the laser beam center in different horizontal planes when the radius of laser beam is 0.5mm, as shown in Figure.3. In Figure.3, when H=14.5mm (H=f_b), the powder effective utilization rate reaches the maximum value, around 5.2%. This is because the powder distribution density in the laser beam center is the largest at this time. Within the range of H<14.5mm, the powder effective utilization rate decreases rapidly with decreasing of H value. When H=7.5mm, the powder effective utilization rate is only 0.2%. Meanwhile, within the range of H>14.5mm, the powder effective utilization rate decreased gradually with increasing of H value, but the reduction rate is significantly lower than that within the range of H<14.5mm. This is well coincident with the variation trend of powder distribution density at the laser beam center in different horizontal planes.

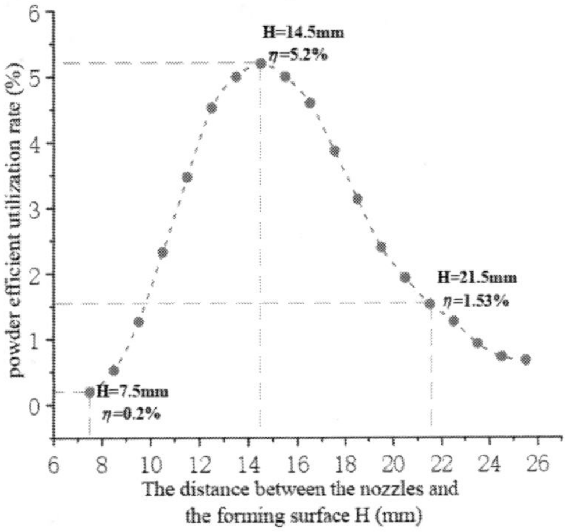

Figure 3. Powder effective utilization rate varies in the center of laser beam in different horizontal planes.

3 Single cladding layer thickness

3.1 Prediction model

In the actual LSF process, the powder materials from powder feeder be sent to the molten pool and solidified into deposition materials. According to the conservation law of powder material mass before and after forming, the single cladding layer thickness can be calculated. For simplicity, we assume that the shape and size of the molten pool are the same as that of the laser beam on the forming surface, and the cross section of the cladding layer is approximately rectangular rather than arched. Based on the measurements of powder effective utilization rate above, the computational formula of single cladding layer thickness can be obtained as follow:

$$\Delta h = \frac{1000 M_p \eta}{60 \rho v B} \qquad (1)$$

where Mp is powder feeding rate (g/min), η is powder effective utilization rate (%), ρ is powder density (g/cm^3), v is laser scanning speed (mm/s), B is laser beam diameter (mm).

From the formula (1), we know that single cladding layer thickness is mainly determined by powder feeding rate, powder effective utilization rate, powder density, laser scanning speed, and the size of laser beam. Powder effective utilization rate and the size of laser beam are constant when the distance between the coaxial powder-feeding nozzles and the forming surface remains unchanged in the LSF process, single cladding layer thickness is mainly affected by powder feeding rate and scanning speed.

3.2 Experimental detection

In order to verify the accuracy of the prediction model of single cladding layer thickness, the focal plane at H=14.5mm was set as the forming surface for the single track experiments. The powder materials and the parameters used in the experiments are listed in Table.1.

Table 1. The powder material and the parameters used in the experiments

Material and Parameters	Values
Powder material	Martensite stainless steel (17-4ph)
Powder density (g/cm^3)	7.8
Powder size (μm)	75~185
Laser power (W)	200
Laser spot (mm)	1.0
Scanning speed (mm/s)	2~12
Powder feeding rate (g/min)	3.0~10.5

When the powder feed rate changes, the comparison results of single cladding layer thickness between theoretical calculation and experimental verification are shown in Figure.4. The top view and three-dimensional diagram of the single cladding layer in the experiments under different powder feed rates are shown in the figure.5. The theoretical calculation results of single cladding layer thickness increased linearly and the experimental results of single cladding layer thickness increased approximately linearly with increasing of powder feed rate. The theoretical calculation results show good agreement with the experimental results. Moreover, it is also found that, the experimental measurement values are slightly lower than the theoretical calculation values.

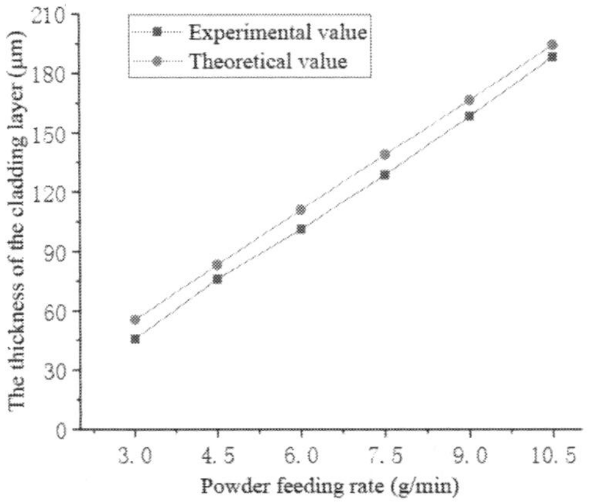

Figure 4. Comparison of the theoretical and experimental value of single cladding layer thickness under different powder feed rate.

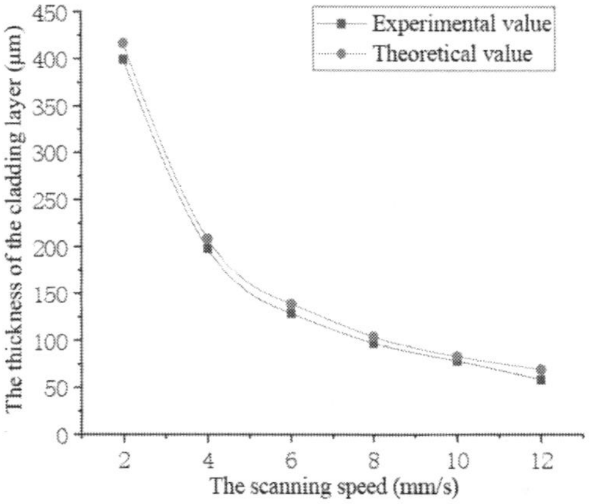

Figure 6. Comparison of the theoretical and experimental value of single cladding layer thickness under different scanning speed.

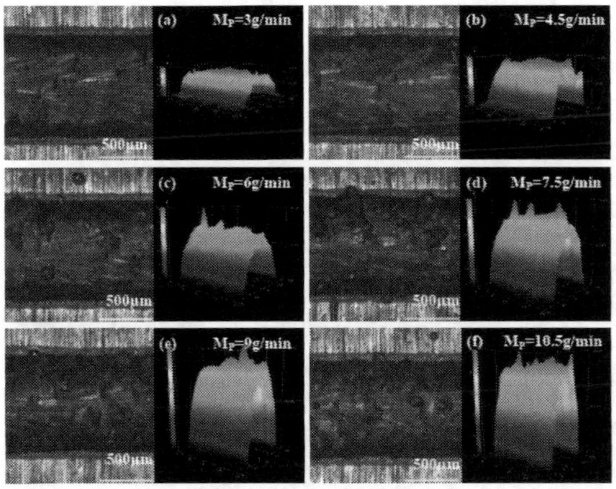

Figure.5 Comparison of single cladding layer morphology under different powder feed rates

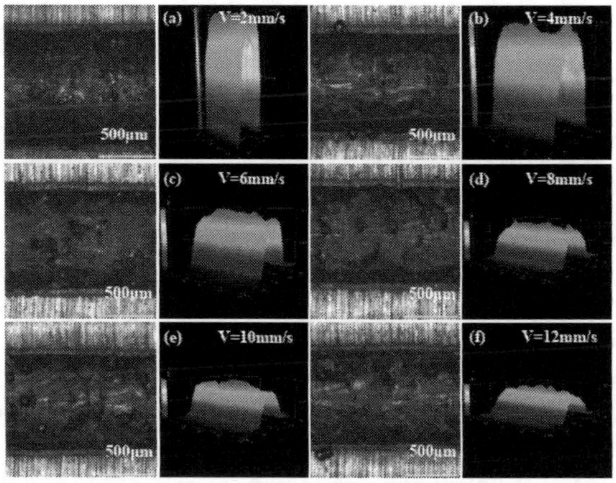

Figure.7 Comparison of single cladding layer morphology under different laser scanning speeds

When the scanning speed changes, the comparison results of single cladding layer thickness between theoretical calculation and experimental verification are shown in Figure.6. The top view and three-dimensional diagram of the single cladding layer in the experiments under different laser scanning speeds are shown in the figure.7. The theoretical calculation results and experimental results of single cladding layer thickness decreased with increasing of scanning speed. The reduction rate of single cladding layer thickness also decreased with increasing of scanning speed. Approximately inversely proportional relationship was established between the single cladding layer thickness and the scanning speed.

4 Analysis and discussion

According to the results above, the theoretical calculation results of single cladding layer thickness are well coincident with the experimental results, and it proves that the measurement method of powder concentration is accurate and reliable. At the same time, it is also found that there are some subtle differences. The experimental measured results are all lower than the theoretical calculation results. By analysing experimental measured results, one possible reason is that the single cladding layer width is inconsistent with the laser spot diameter once the unsuitable laser power are used in the LSF process. In order to verify whether the above conjecture is true, this article investigated the influence of laser power on the width and thickness of single cladding layer via a series of single track experiments. The powder materials and the parameters used in the experiments are listed in Table.2.

Table 2. The powder material and the parameters used in the experiments

Material and Parameters	Values
Powder material	Martensite stainless steel (17-4ph)
Powder density (g/cm^3)	7.8
Powder size (μm)	75~185
Laser power (W)	140~360
Laser spot (mm)	1.0
Scanning speed (mm/s)	6
Powder feeding rate (g/min)	7.5

The experimental results of the width and thickness of single cladding layer under different laser power are shown in Figure.8. As shown in Figure.8, the width and thickness of single cladding layer increase with increasing of laser power. The theoretical calculation value of single cladding layer thickness is 138.9μm under the process parameters selected in the experiments. When the laser power is 260W, the width and thickness of single cladding layer are 1002.6μm and 140.2μm, respectively. This experimental results are highly consistent with the theoretical calculation value. But, the results under other laser power conditions are inconsistent with the theoretical value. Therefore, it can be seen that the accuracy of the theoretical calculation value of single cladding layer can be affected by the laser power. The influence of laser power on the geometry size should be taken into account to improve the prediction accuracy when establishing the prediction model of single cladding layer thickness.

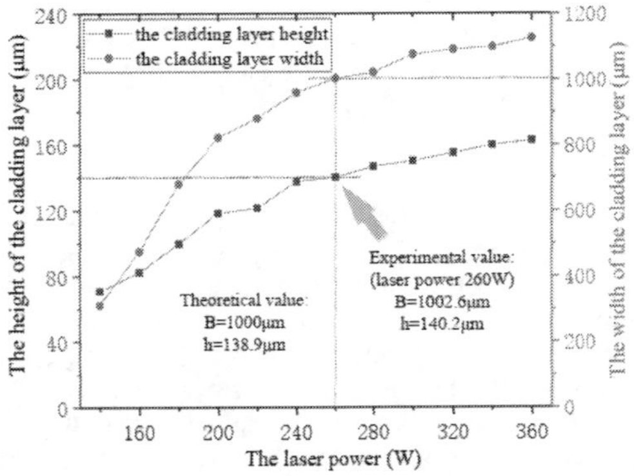

Figure 8. Experimental results of the width and thickness of single cladding layer under different laser power.

5 Conclusions

The main conclusions of this study are as follows:

1) The powder effective utilization rate has a significant impact on the geometry size of single cladding layer during LSF process. Due to the mass of powder materials sent out by the powder feeder per unit time is known, whether the mass of powder materials entering the molten pool can be measured accurately has become the key to measure powder effective utilization rate.

2) This work proposes an alternative method which is based on the weight measurement. Using this method, the powder material entering molten pool was collected by a copper tube with the same size as laser spot, and the correctness of calculating results was verified via a series of single track experiments.

3) The theoretical calculation results of single cladding layer thickness are well coincident with the experimental results, and it proves that the measurement method of powder distribution density is accurate and reliable.

4) Laser power has a great influence on the geometry size of single cladding layer. The influence of laser power on the geometry size should be taken into account when establishing the prediction model of single cladding layer thickness.

Acknowledgements

This work was supported by the National Key R&D Program of China (2018YFB1105900) and the National Natural Science Foundation of China (51575308).

References

1. Tan H. Chen J. Zhang F. Lin X. and Huang W.D 2010 Optics & Laser Technology. 42(1) 47-54.
2. Zhu G.X. Li D.C. Zhang A. and Tang Y. 2011 Optics & Laser Technology. 43(1) 106-113.
3. Gong X.Y. Zhang Y.Z. and Liu M.K. 2013 International Journal of Advanced Manufacturing Technology. 67(9-12) 2501-2509.
4. Pinkerton and Andrew J. 2007 J. of Physics D-Applied Physics. 40(23) 7323-7334.
5. Tabernero I. Lamikiz A. Ukar E. Lopez D. L. Angulo C. and Urbikain G. 2010 J. of Materials Processing Technology. 210(15) 2125-2134.

Analysis of crack mechanism in FGH 98 superalloy during the laser solid forming process

Weisheng Ying [1,2], Shukai Fan [1,2], Fuzhu Han [1,2,*] and Junhua Wang [1,2]

[1] *Department of Mechanical Engineering, Tsinghua University, Beijing 100084, China*
[2] *Beijing Key Lab of Precision/Ultra-precision Manufacturing Equipments and Control, Tsinghua University, Beijing 100084, China*

Abstract. In this study, FGH 98 superalloy parts were fabricated by laser solid forming (LSF) and the microstructure and defects of the LSF part were revealed using optical microscope and scanning electron microscope. It was found that the main defect in the LSF part was cracks. In order to analyze the phase transition process and crack mechanism, the temperature and stress fields were simulated. Based on the simulated result, Ni-Al phase diagram and morphologies of cracks, it can be concluded that the micro segregation caused by the rapid heating and cooling rates was the main reason for the crack defect.

Key words. FGH 98 superalloy; Microstructure; Defects; Crack mechanism; Laser solid forming;

1 Introduction

With the development of aeroengines, the increase of the working temperature demands more superior high-temperature mechanical properties for the material of aeroengine components. FGH 98, a third-generation powder metallurgy nickel-based superalloy, has been widely used for manufacturing aeronautical components due to its good balance between creep resistance, damage tolerance, and tensile properties at elevated temperatures [1]. However, due to the high concentration of refractory elements and the inherent interaction capabilities among the complex precipitation strengthening phases in the FGH 98 superalloy, it is difficult to fabricate components with complex shapes using traditional processing methods [2,3]. In recent years, additive manufacturing (AM) technologies have been developed to produce complex-shaped parts due to its advantages in low production cost, low waste ratio of material and short production cycle. Laser solid forming (LSF), also known as laser metal deposition (LMD) or laser engineered net shaping (LENS), is a typical additive manufacturing technology that combines the advantages of rapid prototyping and laser cladding. Compared to the traditional processing method, LSF can achieve near net shape forming of components without die. Given the notable improvement in production flexibility, LSF is a promising method for manufacturing compact metallic parts with high performance [4].

However, during the rapid heating and cooling process of LSF, there is an inhomogeneous thermal field and steep thermal gradient in the fabricated parts, which leads to a nonequilibrium solidification process and severe stress concentration. Therefore, because of the rapid heating and cooling characteristics, severe defects have a big chance to occur in the LSF part [5]. For FGH 98 superalloy, there are few researches concerning about using LSF to fabricate FGH 98 superalloy parts. The kinds of defects, which may occur in the LSF FGH 98 part, are still unclear. Thus, it is meaningful to have a new attempt to fabricate FGH 98 parts by LSF.

In this study, FGH 98 cuboid samples were fabricated by LSF and defects in the sample were revealed. The temperature field and stress field of the LSF process were simulated by ANSYS software. Based on the characteristics of simulation results and phase diagram of FGH98 superalloy, the microstructure evolution and crack mechanism in the LSF FGH98 superalloy part were explained.

2 Experimental setup

All the LSF experiments involved in this study were performed on a self-developed LSF equipment, which consisted of a SPI 400 W continuous wave laser, a computer numerical control (CNC) working table and a coaxial powder feeding system. FGH 98 spherical powders with particle size in the 45-100 μm range were laser-deposited on a 150 mm × 150 mm × 8 mm stainless steel substrate. Table 1 presents the chemical composition of FGH 98 superalloy. The substrate surface was ground using SiC abrasive paper and cleaned using absolute ethanol in an ultrasonic bath before being laser-deposited. To observe the microstructure of the cross sections, the samples were sectioned using the method of wire electrical discharge machining (WEDM). The metallographic specimens were prepared using standard metallographic techniques, and the etchant used for

* hanfuzhu@mail.tsinghua.edu.cn

https://doi.org/10.1051/matecconf/201824201011

revealing the microstructure consisted of 50 mL H_2O, 50 mL HCl, and 10 g $CuSO_4$.

Table 1. Chemical composition of FGH 98 superalloy (wt %)

Element	FGH 98
Cr	0.05
Co	12.29
Mo	20.68
W	4.61
Al	2.81
Ti	3.32
Ti	3.71
Nb	1.22
Ta	2.47
B	0.02
Zr	0.05
Ni	Bal.

3 Defects in the LSF FGH98 superalloy

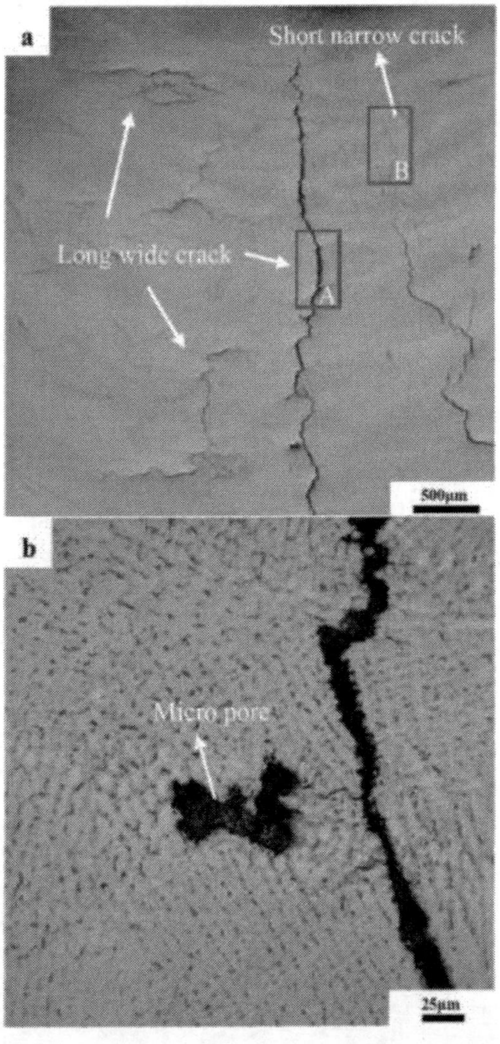

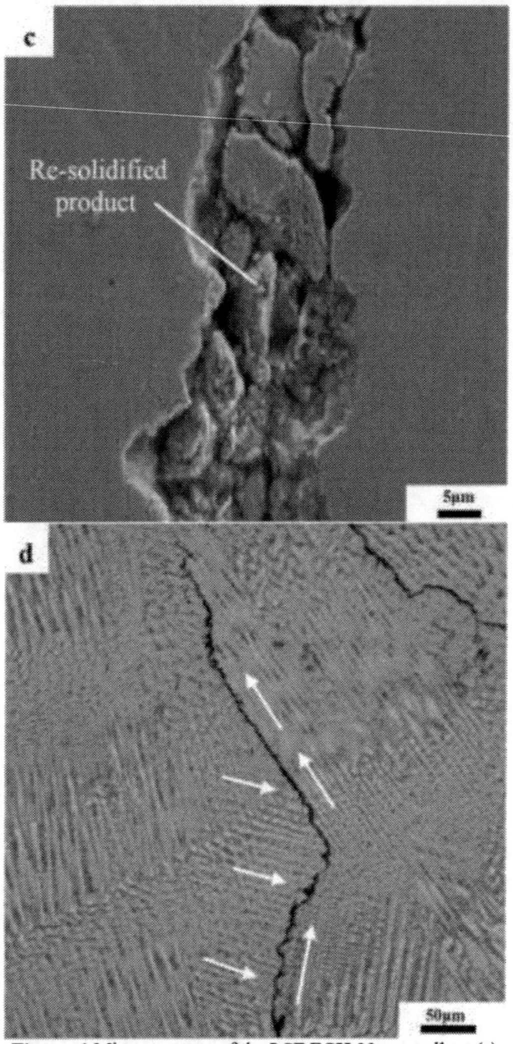

Figure. 1 Microstructure of the LSF FGH 98 superalloy: (a) low magnification OM image, (b) micro pore in the LSF part, (c) highly magnified SEM image of the long wide crack in area A in Figure. 1(a), and (d) highly magnified OM image of the short narrow crack in area B in Figure. 1(a) (the white arrow represents the direction of the grain boundary)

Figure. 1 shows the typical microstructure of the LSF FGH 98 superalloy sample. The microstructure mainly consisted of columnar dendrites which were epitaxially grown from the substrate. Both cracks and micro pores were found on the cross section of the LSF FGH98 part, and the main defect was obviously cracks. Depending on the morphologies of cracks, the cracks can be mainly divided into two types: long wide cracks and short narrow cracks. As shown in Figure. 1(d), it can be found that the short narrow crack initiated in the overlap area between two adjacent scanning passes, where grain boundaries of different directions met, and mainly distributed along grain boundaries. Figure. 1(c) shows the highly magnified SEM images of long cracks. Although the cross direction raster scanning pattern was used in the LSF process, the change of the columnar grain growth direction was not so obvious. Some cracks propagated along the crystallographic orientation of the columnar dendrite and traversed several layers to form a long crack. In the long crack, there were a lot of re-solidified products. Comparing to the short crack, the long crack always had some branches and presented a dendritic morphology.

4 Finite element modeling

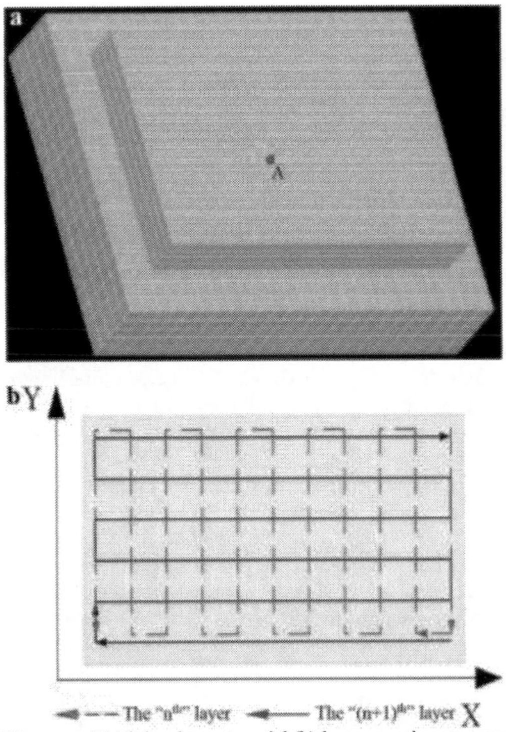

Figure. 2 (a) Finite element model (b) laser scanning strategy

The generation of cracks is decided by the material properties and inner thermal stress. For superalloy, material properties have a close relationship with the temperature. However, it is hard to measure the temperature and stress fields accurately during the LSF process. In order to understand the generation and propagation mechanism of cracks, the temperature and stress fields were simulated by ANSYS software in this study. A 8 mm × 6 mm × 1.2 mm finite element model is established, as shown in Figure. 2. In order to simulate the transient thermal behaviour and subsequent deposition process, the element birth and death technique of ANSYS was adopted. The variations of temperature and stress in the center point of first layer (marked A in Figure. 2) were picked to analyze the temperature and stress fields during the LSF process. The assumptions about material deposition in the model included:

1) The material deposition speed was assumed to be the same as the laser scanning speed.

2) The substrate size was 10 mm × 8 mm × 2 mm, which was far smaller than the substrate size used in the LSF experiments. Thus, the specific heat of substrate was set higher than the normal value.

3) The material properties of FGH 98 were obtained from Chinese Handbook of Aeronautical Materials.

5 Temperature and stress variations during the LSF process

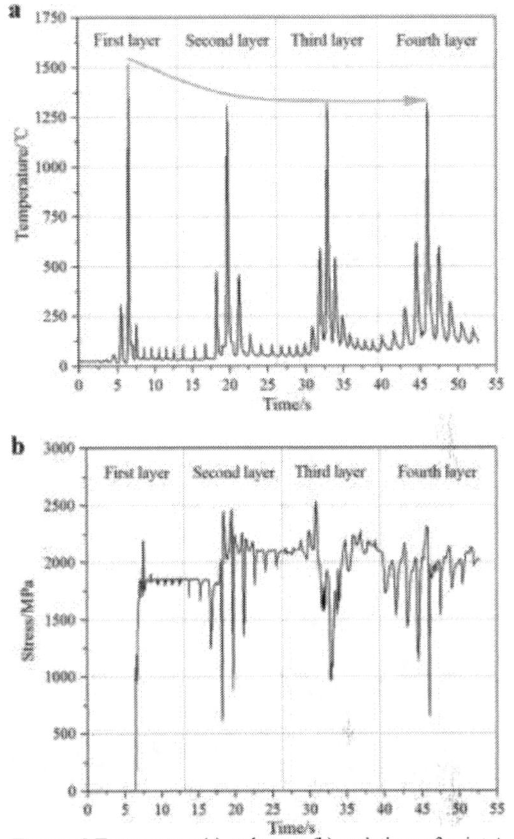

Figure. 3 Temperature (a) and stress (b) variations of point A during the LSF process

Figure. 3 presents the temperature and stress variations of point A during the LSF process. Because of the characteristic of the element birth and death technique, the temperature of point A was low before the element which point A belonged to was activated. Once activated, it can be found that point A experienced a series of periodic temperature cycles. The temperature variation in different layer was similar. The temperature decreased with the increasing distance between point A and the laser spot. Every time when the laser spot passed point A, the temperature reached the highest value during the deposition process of each layer. And the peak temperature in each layer showed a general downward trend with the increase of deposition layers. The temperature variation in Figure. 3 has demonstrated that when the laser spot was close to point A, the temperature would exceed the melting point, which meant the material of point A may be remelted for several times during the LSF process. Moreover, from Figure. 3, it can be found that the cooling rate was also quite high. The fast heating and cooling rate of point A led to a high temperature gradient. Because of the high temperature gradient, there was a huge temperature difference around the laser spot, which led to non-uniform deformation and severe stress concentration. Thus, because of the periodic temperature cycles, point A would also experience a series of periodic stress cycles.

6 Crack mechanism of the LSF FGH98 superalloy

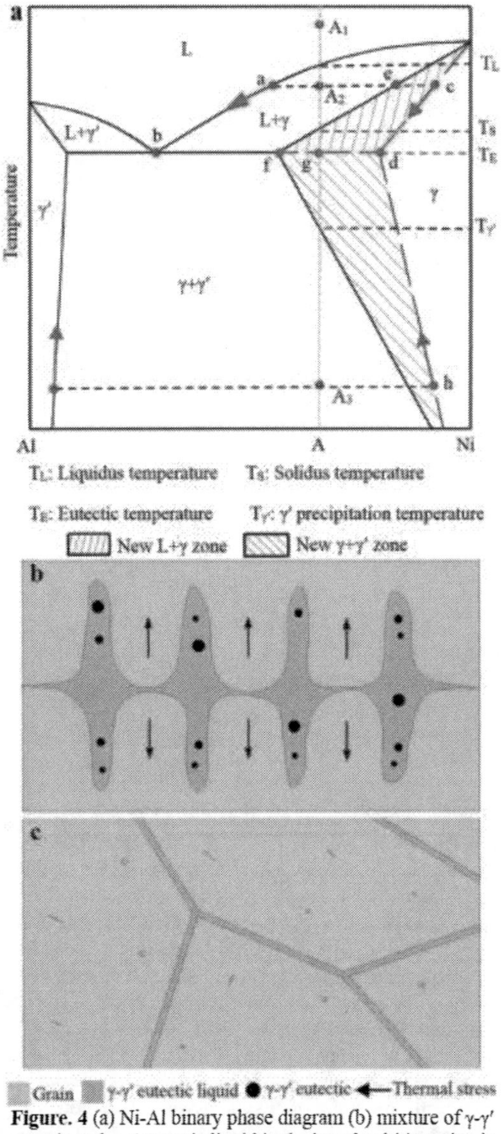

Figure. 4 (a) Ni-Al binary phase diagram (b) mixture of γ-γ′ eutectic and γ-γ′ eutectic liquid in the interdendritic region in the rapid solidification process (c) low-melting point eutectic liquid film along the grain boundary in the reheating process

The crack is mainly caused by the high concentration of Al+Ti in superalloy, and the phase transformation behaviors in the Ni-Al or Ni-Ti system are similar [6,7]. Thus, to understand the effect of temperature cycle on material properties, a Ni-Al binary phase diagram was used to analyze the phase transition process. For FGH 98 superalloy, the chemical composition was close to point A in the phase diagram. When FGH 98 superalloy started to cool down from point A1 or A2, based on the phase diagram it can be found that there would be no γ′ phase under slow cooling rate until the temperature was decreased to the γ′ precipitation temperature. However, according to the simulation result, the cooling rate was very high. During the solidification process, there would be not enough time for Al element to diffuse in the γ matrix, which led to a higher Al content in the liquid. Without considering the uneven diffusion of Al element in the liquid, the solid component changed following the line cd instead of line ef in the phase diagram while the liquid component still changed with the line ab. When the temperature was decreased to the eutectic temperature, according to the lever rule and new phase diagram, there would be still some liquid in the interdendritic regions, and the liquid component was close to the eutectic point b. Thus, during the rapid solidification process, there was a high possibility for the eutectic reaction to occur. As shown in Figure. 4(b), a mixture of γ-γ′ eutectic and γ-γ′ eutectic liquid might exist in the interdendritic region, which resulted in the decrease in ductility. When the thermal stress overcame the low ductility of the material, the mixture of γ-γ′ eutectic and γ-γ′ eutectic liquid in the interdendritic region would be torn, causing solidification cracks along the interdendritic regions. According to the analysis above, we thought the long crack may be caused by the rapid solidification process and belong to the solidification crack. The re-solidified product in the long crack was examined by EDS. Based on the EDS results, it can be found that the composition of re-solidified product corresponded with the γ-γ′ eutectic, which proved the analysis above.

Table 2. EDS results of the re-solidified product and surrounding material (wt %)

Element	Re-solidified product	Surrounding material
Cr	8.31	12.32
Co	17.69	20.53
Mo	9.23	4.71
W	3.03	3.11
Al	7.65	3.40
Ti	10.86	3.69
Nb	2.20	0.72
Ta	3.66	2.47
Ni	37.38	49.06

For the solidified material (such as A3 in the phase diagram), during the rapid reheating process, when the reheating temperature was higher than the eutectic temperature but lower than the solidus temperature, based on the phase diagram, the precipitated γ′ phase would generally dissolve into the γ matrix under slow cooling rate. And there would be no liquid during the reheating process. However, the γ′ phase usually have no sufficient time to dissolve completely into the matrix. The actual solidus curve was the blue line hd in the phase diagram. As shown in Figure. 4(c), the remaining γ' phase near the grain boundaries would react with the γ matrix to produce a continuous low-melting point eutectic liquid film along the grain boundary when the previous deposit was reheated to the eutectic reaction temperature. The low melting point eutectic film weakened the grain boundary cohesion and led to a decrease in the ductility of the LSF part. Therefore, when the thermal stress exceeded the critical tensile stress, the liquation film would be undrawn, causing cracks along the grain boundary. And this analysis correlated well with the characteristics of narrow cracks in Figure. 1.

7 Conclusions

The main conclusions of this study are as follows:

1) FGH 98 superalloy parts were fabricated by LSF. It was found that the main defect in the LSF FGH 98 part was cracks, and two kinds of cracks were found on the cross section.

2) The temperature and stress fields were simulated. The simulation result has demonstrated that the LSF part experienced a series of periodic temperature and stress cycles.

3) The crack mechanism during the LSF process was explained according to the phase diagram and simulation results. It can be concluded that the micro segregation caused by the rapid heating and cooling rates was the main reason for the crack defect.

Acknowledgements

This work was supported by the National Key R&D Program of China (2018YFB1105900) and the National Natural Science Foundation of China (51575308).

References

1. R.C. Reed, *Superalloys: fundamentals and applications* (Cambridge University Press, Cambridge, 2006)
2. E.O. Ezugwu, Int. J. Mach. Tools Manuf. **45**, 1353-1367 (2005)
3. D. Zhu, X. Zhang, H. Ding, Int. J. Mach. Tools Manuf. **64**, 60-77 (2013).
4. W.D. Huang, X. Lin, J. Chen, Z.X. Liu, Y.M. Li, *Laser solid forming technology* (Northwestern Polytechnical University Press, Xi'an, 2007)
5. M. Alimardani, V. Fallah, M. Iravani-Tabrizipour, A. Khajepour, J Mater Process Technol. **212**, 113–119 (2012)
6. R. Kainuma, M. Ichinose, I. Ohnuma, K. Ishida, Mater Sci Eng A. **312**, 168–175 (2001)
7. K. Otsuka, X. Ren, Prog Mater Sci. **50**, 511–678 (2005)

Design and Simulation Analysis of Micro-Porous Structure in Infiltration-Type Self-Lubrication Composites Matrix

HAN Ying [1,a]

[1] *School of Mechanical Engineering, Shandong Yingcai University, Ji'nan, 250104, China*

Abstract. Metal ceramic preforms were fabricated by vacuum sintering method in this work. Ti-Fe-Cr-W-Mo-V powder was used as the basic material, TiH2 and CaCO3 were used as the composite pore-forming material and Al2O3 was used as dispersal particles. The microstructure and the worn surface morphology of the preforms and the composites were analyzed using scanning electron microscope (SEM) and energy dispersive spectrometer (EDS). The experimental results showed that when the micro-pore shapes are circular and quadrilateral respectively, with the increasing load, the Mises equivalent stress distribution generated by the single cell model is quite different, but it has symmetry, and the maximum stress appears around the interpenetrating micro-pore. Using FeCrWMoV alloy powder as the metal phase and TiC powder as the ceramic phase, the micro-porous matrix was prepared by powder metallurgy sintering with the addition of a certain volume fraction of compound pore-making agent.

1 Introduction

The new type of infiltration self-lubrication composites is prepared by using bionic tribology. Under vacuum pressure, the melt solid lubricant is immersed in the micro- pore matrix. Under the action of high temperature and friction heat, the solid lubricant is precipitated out to the friction interface along the micro-pore structure[1]. The material has good mechanical properties and self-lubricating properties under special working conditions such as high temperature and high pressure, and is widely used in aerospace, electronic machinery and other industries.

The main function of the micro-pore matrix is to bear the load, which affects the mechanical properties and various properties of composite materials[2,3]. How to control the micro-pore structure of the matrix, so that it can store enough lubricant to meet the requirements of lubrication performance[4], and at the same time, it has enough strength to meet the requirements of mechanical performance is the key of this paper.

2 Finite element simulation

2.1 Pore shape

The micro-pore structure of the infiltration self-lubrication composites matrix is complex, which mainly depends on the powder particles and the production process. Micro-pore structure characteristics affect the mechanical properties of the self-lubricating composites, which is the key to determine the number of infiltrated solid lubricant, the uniformity of lubricant distribution and whether the lubrication film can be formed at the friction interface[5]. Therefore, the micro-pore shape is an important structural characteristic parameter of porous matrix.

The micro-pore matrix was prepared by powder metallurgy sintering, and its internal structure was very complex. In order to facilitate the analysis, the micro-pore structure is assumed as follows: the micro-pore is uniformly distributed; the micro-pore size is equal; the wall thickness is uniform and isotropic. Taking the regular quadrilateral as the basic cell, the structural sections of different micro-pore shapes are shown in figure1.

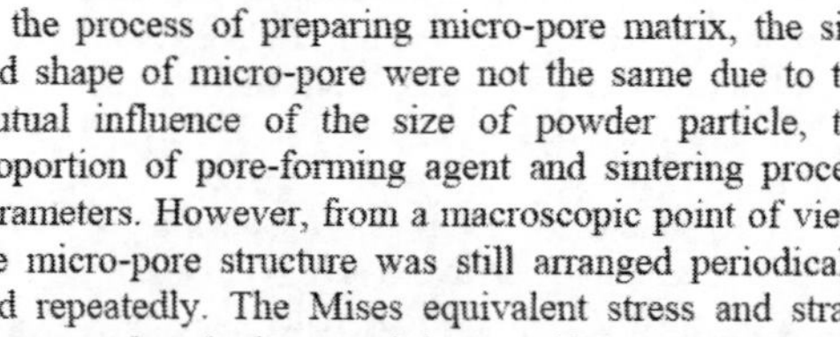

Figure 1. Micropore section diagram: (a) circular (b)quadrangle

2.2 Finite element simulation

In the process of preparing micro-pore matrix, the size and shape of micro-pore were not the same due to the mutual influence of the size of powder particle, the proportion of pore-forming agent and sintering process parameters. However, from a macroscopic point of view, the micro-pore structure was still arranged periodically and repeatedly. The Mises equivalent stress and strain were analyzed by applying uniformly distributed displacement load to a three-dimensional single cell model in space.

[a] Corresponding author: abchanying@126.com

Figure 2 is the equivalent stress distribution cloud map produced by the cellular model with a porosity of 20%. It can be seen from the figure that different micro-pore structure models produce different degrees of deformation and different stress values, but the maximum stress appears around the interpenetrating micro-pore. With the increase of external load, the value of equivalent stress increased gradually. When the loading is the same, the Mises equivalent stress produced by the circular microcellular model is the smallest, followed by the quadrangle.

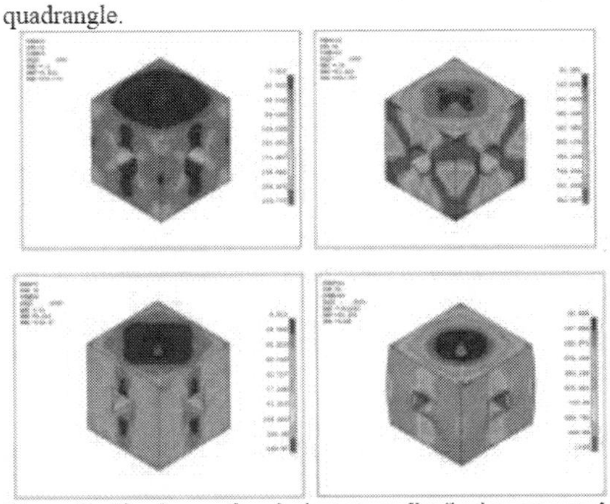

Figure 2. Cloud map of equivalent stress distribution generated by the single cell model

3 Experimental procedure

3.1 Preparation of sample

The matrix powder was mixed with 80%(volume fraction) in Ti-Fe-Cr-W-Mo-V(the chemical composition as: 20%Ti, 4%Cr, 6%W, 5%Mo, 3%V, Fe in balance) alloy powders (particle size of 15 to 25μm) and 20%(volume fraction)Al_2O_3 ceramic powder (particle size of 20 to 25μm). And 40%(volume fraction)of TiH_2 and 60%(volume fraction) of $CaCO_3$ were added as complex pore-forming agent. It was ball-milled for 4h, and then die-pressed into cylindrical specimens of Ø12mm×16mm under a 600MPa of compaction pressure. After compaction, sample was vacuum sintered under 0.1 to 0.01Pa and sintering temperature was 1200°C to 1300°C.

3.2 Experimental method

The liquid static equilibrium method is used to test the density, the porosity and the volume of infiltration lubricant of the preform. The micro hardness and crushing strength of samples were tested using HVS-type micro hardness tester and YE-600 type hydraulic testing machine respectively. Friction and wear properties of the composites were investigated using a high temperature XP-5 pin-on-disk wear tester. The pin in the form of 12 mm diameter and 15 mm length was rounded to hemispherical type of 10 mm diameter at one end and surface roughness of 0.2 μm to 0.8 μm. The disk was made of Ti-Al ceramic coating materials with geometric size of 55 mm×5 mm. The hardness of the disk was HRC 62 to 67 and surface roughness of 0.2μm. The friction and wear tester were conducted at temperature of 600°C and sliding speed of 0.139 m/s. The samples microstructures and worn surface morphology were characterized using scanning electron microscope (SEM) equipped with an energy dispersive spectroscope (EDS).

4.Results and discussion

4.1 Microstructure

Figure 3 is matrix powder (metal and ceramic phase volume fraction ratio of 80:20) particle size is 55-75 microns, compound pore-forming agent content is 7%, at 1230 ℃ sintering, thermal insulation, after 60 min sintered sample pore structure characteristics of different cross section. As can be seen from figure 3(a), the micro-pore distribution on the surface of the sample is relatively uniform, the pore size is basically the same, and the micro-pore shape is mainly round, square and irregular, etc. As can be seen from figure 3(b), the internal structure of the sample is intersected into a network, similar to the structure of human sweat glands. Fig. 3(c) and (d) are enlarged SEM images with square and circular micro-pore shapes respectively. In the process of actual preparation, the matrix of self-lubricating composite material with thick wall cell structure was obtained by taking metal ceramics as the raw material and using compound hole making agent through secondary hole making. Because of the different sintering process parameters such as sintering temperature, heating speed and heat preservation time, various pore shapes will be formed simultaneously in the sintering matrix, and the porosity percentage of various pore shapes will be different.

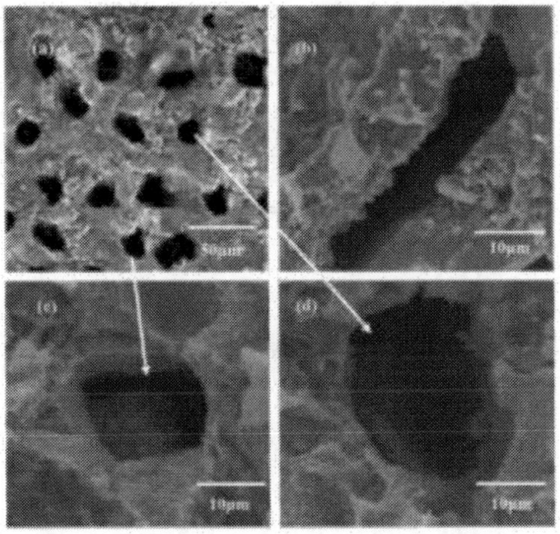

Figure 3. Micro-pore structure of sintered sample: (a) surface; (b) axial section; (c) square hole; (d) round hole

4.2 The determination of sintering process parameters

Fig.4 shows the change curve of metal ceramic sintered body about micro hardness and crushing strength. With the increase of sintering temperature, the micro hardness and crushing strength of preform was improved. When the temperature was more than 1230°C, the curve became flat. This situation is not in conformity with the regular pattern that intensity was dropped according to the index law with the increase of porosity [6]. It was suggested that the preform have corresponding porosity under a certain sintering temperature. With the increase of sintering temperature, the preform tended to be dense and the porosity was decreased. The diffusion speed was accelerated and the activity of powder was enhanced with the increasing of sintering temperature [7, 8]. In consideration of the requirements of pore structure and wear resistance about the preform, the advisable sintering temperature is about1230°C.

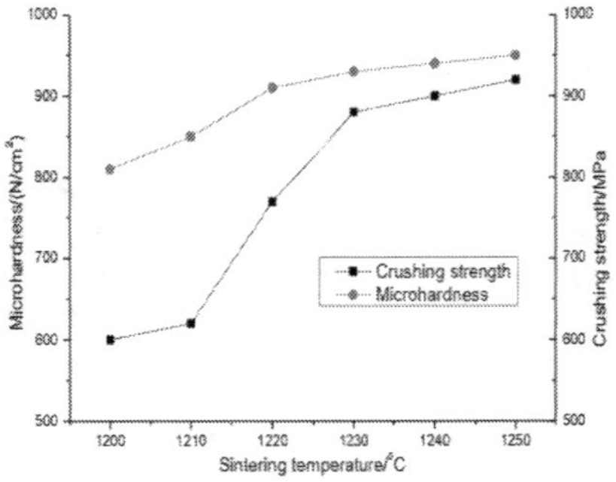

Figure 4. Influence of sintering temperature on micro hardness and crushing strength of the preform

5 Conclusion

When the micro-pore shapes are circular and quadrilateral respectively, with the increasing load, the Mises equivalent stress distribution generated by the single cell model is quite different, but it has symmetry, and the maximum stress appears around the interpenetrating micro-pore. Using FeCrWMoV alloy powder as the metal phase and TiC powder as the ceramic phase, the micro-porous matrix was prepared by powder metallurgy sintering with the addition of a certain volume fraction of compound pore-making agent. The micro-pore shape is regular (mainly round or square), and the pores are evenly distributed and interspersed into networks。

References

1. J. W. Qiu, Y. Liu, Y. B. Liu: Material & Design. Vol. 33(2012)
2. X. L. Sun, Y. Liu, Y. Lu: Powder Metallurgy Technology. Vol. 19(2001)
3. Z. M. Liu: Wear. Vol. 262(2007)
4. J. M. Xue, C.X. Sun: Journal of East China University of Science and Technology, Vol. 28(2002)
5. Y. J. Wang, Z .M .Liu: Chinese Journal of Mechanical Engineering.Vol.42.(2006)
6. H. D. Ding: The Transaction of Nonferrous Metals Society of China, Vol. 11(2001)
7. H.Chang, J. Binner, R. Higginson, P. Myers, P. Webb, G.King: Materials Science and Engineering A. Vol. 529(2011)
8. Z. F. Xu, H. Yu: Chinese journal of nonferrous metals, Vol. 169(2006)

The priority classification method of mobile users under direct power transaction

Wang Lei[1,2], Gao Chuncheng[1,2], Zhang Qian[1,2,a] and Dai Yong[1,2]

[1] Nari Group Corporation/State Grid Electric Power Research Institute Nanjing China

[2] Beijing Kedong Electric Power Control System Co., Ltd. Beijing China

Abstract. With the development of mobile internet technology, market-oriented electricity reform forms a new transaction mode that "Internet + power trading", which further promotes the liberalization of retail side market. A large number of mobile power users actively participate in the electricity market transaction. However, the current standards for mobile users who are accessed to the market are qualitative standards, lacking of quantitative analysis. The priority classification method of mobile users under direct power transaction is proposed in the paper. First, the priority evaluation index set of mobile users is established from four dimensions that credit condition, environmental condition, power consumption condition and load importance to achieve evaluation quantitatively. Secondly, K-means clustering algorithm based on SOM is used to cluster mobile users. Finally, the paper recognizes the various types of mobile users as a whole and estimates their priorities using the improved AHP(Analytic Hierarchy Process). The method realizes the specific division of mobile users from a point of quantitative view, and solves the problem of only qualitative standard currently, to optimize the development of direct power transaction and promote the marketization of power market.

1 Introductuion

The direct power transaction is that power plants and power users determine the purchase of electricity and electricity prices through the form of direct power transactions, and then commission power grid companies to transfer the power of the agreement from the power generation enterprises to the power users and pay for the transmission and distribution services undertaken by the grid companies.

Meanwhile, with the continuous deepening of the mobile direct power transaction, a large number of mobile power users actively participate in the direct power transaction. The domestic and foreign electricity market has defined the conditions for the transfer of members of the direct electricity market. All electricity users in the US electricity market can trade directly online, but large users need to pay for grid service fees to power grid companies. The European electricity market is open to users whose electricity consumption is more than 9 million kWh. At present, large domestic direct purchase of electricity to emphasize too much access to large users of electricity scale, although it is clearly stipulated in the transaction implementation in various regions that large users involved in direct power transactions should conform to the national industrial policies with low energy consumption per unit output in the same industry and small pollution emission.

However, the admittance standards are all qualitative, and in practice it will involve how to make a concrete division. In order to realize the quantitative evaluation of the priority [5-6] of the mobile terminal users, this paper proposes a priority evaluation index system [7-8], evaluates its priority by using improved AHP [9-10], which achieves the quantitative division of mobile users.

2 The comprehensive evaluation model of supervision risk in electricity transaction based on unascertained rational number

The mobile user priority evaluation indicator system includes four first-level indicators which are credit condition, environmental condition, power consumption condition and load importance, and 11 second-level indicators, as shown in the following figure.

[a] Corresponding author: E-mail:hbkjdx2009@126.com

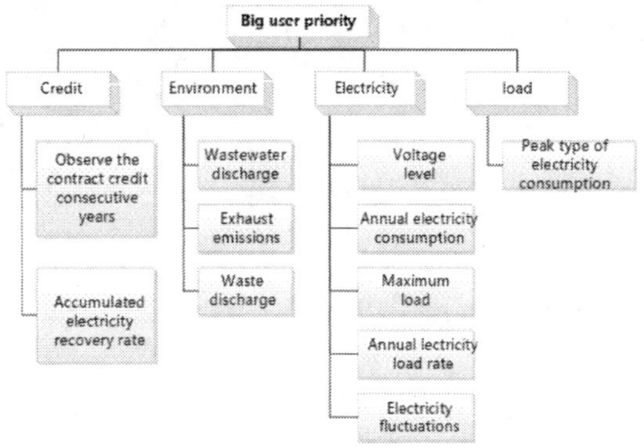

Figure 1. evaluation indicator system

(1) Credit condition （S1）

Accumulated electricity recovery rate (S12) is as follow. Refer to "(1),"

$$S12 = \frac{a}{b} \tag{1}$$

In the formula:

a: a user's received electricity bill over the past three years.

b: a user's receivable electricity bill over the past three years.

(2)Environmental condition （S2）

Environmental condition （S2） is as follow. Refer to "(2),"

$$S2 = \frac{1}{2}\sum_{i=1}^{3} S2i \tag{2}$$

The expression of each sub-indicator is as follow. Refer to "(3),"

$$S2i = \begin{cases} 0 \\ (S_i - S_i^-)/(S_i^+ - S_i) \\ 1 \end{cases} \tag{3}$$

In the formula:

Si: the i-th pollutant emissions， i=1, 2, 3.

S_i^+ and S_i^- respectively represent the upper and lower limits of such pollutant discharge stipulated by our country.

(3) Electricity condition （S3）

The users' voltage class is generally divided into five categories, that is, less than 1 KV, 1-10 KV, 35-110 KV, 110 KV, and 220 KV and above. Accordingly, the quantized values of the user's voltage level (S31) are respectively recorded as 1, 2, 3, 4, and 5.

Annual electric load rate (S34) is expressed as follow. Refer to "(4),"

$$S34 = \frac{10000 * S32}{8760 * S33} \tag{4}$$

Electricity fluctuations (S35)is expressed as follow. Refer to "(5)-(6),"

$$S35 = \frac{\sqrt{\frac{1}{12-1}\sum_{i=1}^{12}(L_i - L)^2}}{L} \tag{5}$$

$$L = \frac{1}{12}\sum_{i=1}^{12} L_i \tag{6}$$

In the formula:

L (ten thousand kWh): the average of the monthly electricity consumption.

Li (ten thousand kWh): the electricity consumption of the big user in the I–th month. S5 is the standard deviation of the large user's monthly electricity consumption divided by the average monthly electricity consumption, which indicates the fluctuation and periodicity of the user's monthly electricity consumption.

(4)Load condition （S4）

Peak-class enterprises have excellent power consumption insurance category, A class, B class, C class and limited class. Accordingly, Peak type of electricity consumption (S41) are respectively recorded as 5, 4, 3, 2 and 1.

3 Clustering Analysis of Mobile Terminal Users Based on Improved K-means Clustering Algorithm

K-means clustering algorithm based on SOM belongs to two-stage calculation method. In the first stage of the initial clustering, SOM first cluster massive data samples and classify the eigenvectors with similar characteristics to the same category, so that the sample data can be clustered into different categories, and the number of categories and the central point of each category can be obtained. In the second stage, the K-means clustering uses the result of the first stage as the initial value input and further clusters to form the final clustering result.

4 The Priority Evaluation of Mobile Terminal Users Based on Improved AHP Algorithm

(1) The improved AHP algorithm

The traditional AHP usually uses the I-9 scale method to establish the judgment matrix. If it is not consistent when tested, the matrix will affect the effect of AHP's optimal sorting, and it must be reconstructed until it passes the test, so the computation is large and the accuracy is not high.

The improved AHP adopts a new scale method - the three-scale method, which does not need consistency test when judging and comparing the relative importance of the factors to each other. Therefore, the number of iterations can be greatly reduced, the convergence speed can be improved and the calculation accuracy can be met. The comparison matrix constructed by the three standard method is expressed as follow. Refer to "(7),"

$$a_{ij} = \begin{cases} 2: \textit{Index i ismore important thanindex j} \\ 1: \textit{Index i is as important as the index j} \\ 0: \textit{Index i less important than index j} \end{cases} \tag{7}$$

In the formula,

a_{ij}: the element value of the matrix A

Besides, calculate the importance ranking index r_i of each index. Construct the judgment matrix B and find the quasi-optimal consistent matrix D.

(2) Mobile terminal user priority evaluation

TOPSIS does not reflect the subjective preference in the evaluation. In this paper, TOPSIS is used to determine the final solution.

Calculate the approximate degree T_i of each sample and the virtual result. Refer to "(8),"

$$T_i = d_i^- / (d_i^+ + d_i^-) \quad (8)$$

T_i is between 0 and 1. The samples are sorted according to T_i, that is, the greater the T_i shows, the closer the solution represented by the result is to the optimal solution and the farther it is from the worst solution.

5 Conclusion

In this paper, the priority classification method of mobile users under direct power transaction is proposed. The method selects evaluation indexes and establishes the mobile user priority evaluation indicator system, which includes four aspects, being credit condition, environmental condition, power consumption condition and load importance, and quantifies the priority of mobile users from multiple indicators, solving the problem of now only qualitative criteria, and laying a research foundation for mobile terminal user priority classification.

References

[1] Qi-ping FENG. "Research on Electricity Transmission-Distribution Independent and Large Consumers Direct Purchasing Based on Power Market," Central South University, 2014.

[2] Liang TAO. "Research on the Direct Power Purchase for the Large Customer," Beijing: North China Electric Power University, 2014.

[3] Fu-bin LIU, Tao PENG, Xian ZHANG, Bo PANG, Ye HAN. "Pilot and Promotion Recommendations for Direct Trading between Power Users and Plants," East China Electric Power, vol.41, no.9, pp.1941-1944, 2013.

[4] Wei CHEN, Jing-jing YANG. "'Palm Power' bloom capital," State Grid, no.9, pp.86-87, 2015.

[5] Qi-zong SUN, Jia-ji ZHONG, Chuan CHEN, Er-tian HUA. "Method of determining customized priority based on customer preferences," Journal of Mechanical & Electrical Engineering, vol.34, no.11, pp.1354-1358, 2017.

[6] Li-qiong JIANG. "Research on Access Technologies of Dynamic Users in Space-based Network on Account of Traffic Priority," Harbin Institute of Technology, 2017.

[7] Yang-sen OU, Li-yuan LIU. "Reliability Index System of Distribution Network for Power Consumer and Its Comprehensive Assessment Method Power," System Technology, vol.41, no.1, pp.215-222, 2017.

[8] OU Yangsen, Shi Yili. "A new improved entropy method and its application in power quality evaluation," Automation of Electric Power Systems, vol.37, no. 21, pp.156-159+164, 2013.

[9] Tong LIU, Mu HE, Yu-xiao WANG, Qing-yu YAN. "Comprehensive Evaluation on the Satisfaction for Water Park in Tianjin with the AHP and Kano Model," Journal of Shandong Agricultural University (Natural Science Edition), vol.47, no.3, pp.417-424, 2016.

[10] Yao-tian ZHANG, Xu-cheng ZHANG, Ming-shun JIA, Xiang-shang XUE. "Adaptive evaluation method based on analytic hierarchy process," Journal of Beijing University of Aeronautics and Astronautics, vol.42, no.5, pp.1065-1070, 2016.

Acknowledgment

This work was funded by the headquarters technology project of state grid corporation of China "Research on the operation risk analysis and support technology of electricity market in the interaction of supply and demand".

Seismic retrofit of historical masonry walls by means of natural basalt fiber strips system (BFRP). Pre-modern concepts and innovative materials

Giacomo Tempesta[1,a]

[1]*Department of Architecture, University of Florence, Italy*

Abstract. The paper deals with the design approach for post-seismic interventions aimed at safeguarding historical buildings in seismic areas, consisting prevalently of masonry structures that, by type of technology, even if within exceptional cases respectful of "the rule of art", do not guarantee the compliance with the resources necessary for the demand for resilient and structural capacity. In particular, the paper deals with the issue of applying technologies that use constructive ideas and concepts of the past or, as they are usually called pre-modern, reinterpreting them through the use of innovative and extremely performing materials.

1 Introduction

Going through the historical constructive solutions developed and used in past epochs to give an answer to the anti-seismic design problem, we note a great homogeneity from the point of view of the choice of materials and techniques used. Focusing on the Mediterranean Basin, the propensity for the use of mixed timber-masonry systems has been ascertained. This choice should not be surprising, since it was dictated by the mechanical features of timber that provide masonry walls with the capacity of withstanding the shear and bending actions provoked by an earthquake acting along the wall plan; a very limited capacity in unreinforced masonry. Already in Roman times timber members inserted, in a targeted way, within masonry walls were a widespread technique. Vitruvius himself [1] suggested the use of "*travicelli d'ulivo brustolato per tutta la grossezza del muro*" ("burnt Olive members within the entire thickness of the wall") to provide a link in such a way as "*entrambe le fronti del muro [...] abbiano perpetua durata*" ("both facings of the wall [...] can live forever"). Although the spread of timber-masonry techniques cannot be attributed only to the need of building anti-seismic constructions, most of these techniques were used in highly seismic areas of the entire Mediterranean Basin and that some specific constrictive systems were developed as a consequence of the occurrence of important seismic events.

Archaeological excavations carried out in the sites of Pompeii [2] and Herculaneum [3] in the 19th century highlighted that the constructive technique based on the use of mixed timber-masonry, called *Opus Craticium* (Fig. 1a), was based on the use of a timber frame that divides the masonry wall in smaller portions, constituted by transversal, vertical and bracing members joined together.

According to Bianchini [4], it is a building system that seems to be became established in Italy since the 6th century and that, in a later time, continued to be used as an alternative to the stronger *strcturae caementiciae*, in rural buildings as well as in city dwellings, at the upper floors and to build internal partitions.

Even if it is much less known, the masonry "alla beneventana" is an interesting technique that anticipates the Bourbon "casa baraccata". It is referred to as a preferred construction technique for reconstruction after the Gargano earthquake occurred in 1627.

Its name being taken from the city of Benevento, it is constituted by a load-bearing timber frame with infill masonry: a set of posts fixed on a stone base and an infill of lightweight materials (such as wicker or reeds linked to chestnut laths), covered with mortar.

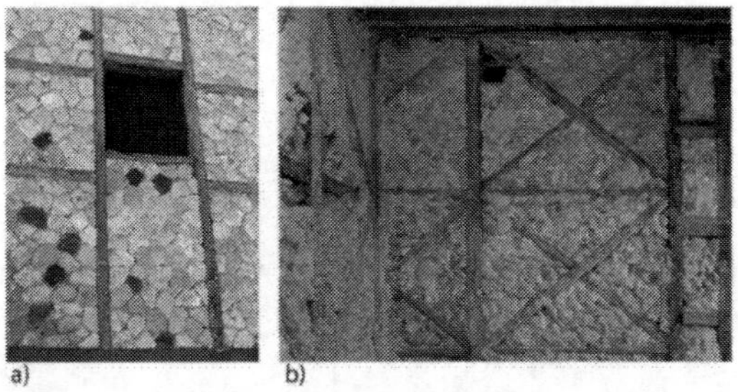

a) b)

Figure 1. a) *Opus Craticium*; b) Gaiola Pombalina, 1775.

Thanks to the several surveys in the site of Irpina, carried out immediately after the strong seismic events occurred in 1980, it was ascertained that the structural response of the masonry "alla beneventana" was

[a] Corresponding author: giacomo.tempesta@unifi.it

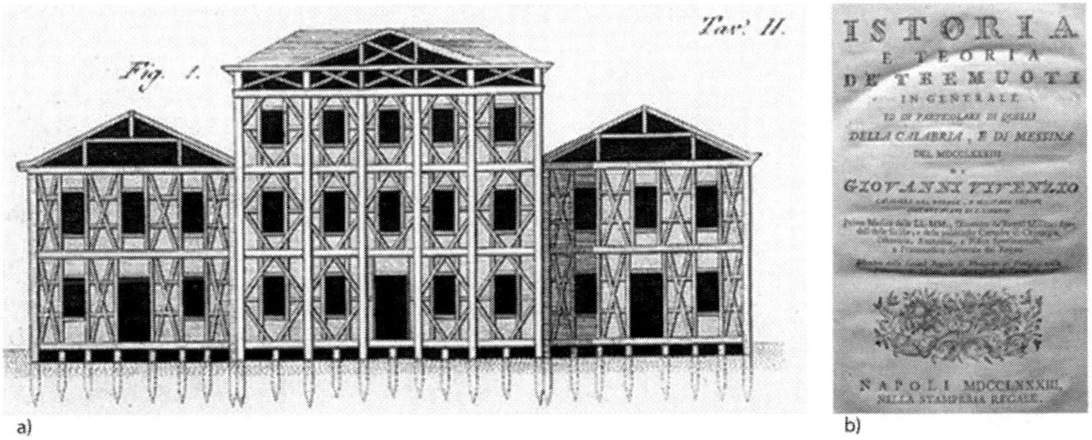

a) b)

Figure 2. Casa Baraccata: a) the sketch and b) the treatise by Giovanni Vivenzio, 1783.

performing and technologically functioning. Indeed, its success was due to the result obtained, consisting in partitioning walls in small portions, separated by wooden rods linked to the principal members, in which crack formation is allowed. Sometimes, some portions of the wall may also collapse, but the area of failure is reduced and is confined within the timber partitions.

Similar solutions were adopted in Turkey [5-6]: it's the case of the "hatul" and "himis" constructions, that resisted the earthquake in 1999. This constructive technique, of Ottoman origin, provides the building with a timber frame structure composed of vertical, horizontal and inclined members, filled with raw earth herring bone bond masonry. The "Dhajji dewari" is another timber frame with stone and earth infill, typically used in the mountain regions of South Asia.

"Gaiola", also called "Gaiola Pombalina" (Fig. 1b), is the building system conceived by Engineer Pombal for the post-seismic reconstruction of the Baixa area in Lisbon, immediately after the high magnitude earthquake occurred in 1775. Also this technique provides the use of a timber frame composed of posts and beams arranged in such a way as to form rather square portions. St. Andrew's crosses are inserted within each partition, with a brace function, together with an infill made of stone and brick masonry, usually recovered from rubble [7]. Masonry infill protrudes slightly from the timber frame, so that the positioning of a layer of covering plaster is allowed. At the end of the eighteenth century, the mixed timber-masonry constructive system, that is more appropriate to satisfy the anti-seismic requirements, is spreading. This technique derives from a fairly widespread tradition in timber constructions built in the Calabria region and filtered from the Enlightenment school of thinking that characterizes the era of the Bourbon domination in southern Italy.

Giovanni Vivenzio himself, doctor of the Royal Household and university professor, states that his earthquake resistant prototype, called "Case formate di legno (Houses built using timber members)" was deduced by the "[..]...osservazioni fatte sul luogo medesimo... (on-site inspections...)" such as, for example, "... Filogaso, l'antico Palazzo del Conte di Nocera ... costrutto nel passato secolo, di legno, e rivestito solamente di fabbrica, si rimase in tutto l'interiore illeso, nel mentre che il Paese fu uguagliato al suolo ... (... Filogaso, the ancient palace of the Count of Nocera ... built in the previous century using timber members and covered with masonry, that did not suffer damage inside, while all other buildings collapsed)" (Fig. 2). This system, denoted as "casa baraccata" or Bourbon anti-seismic system, was conceived by Engineer Ignazio La Vega after the Calabrian earthquake occurred in 1783. La Vega developed and improved this system on the basis of the knowledge and observations derived from the direction of the archaeological excavations of Pompeii and Herculaneum as well as from those obtained from examples already present in the traditional Calabrian building, such as the Bishop's building in Mileto.

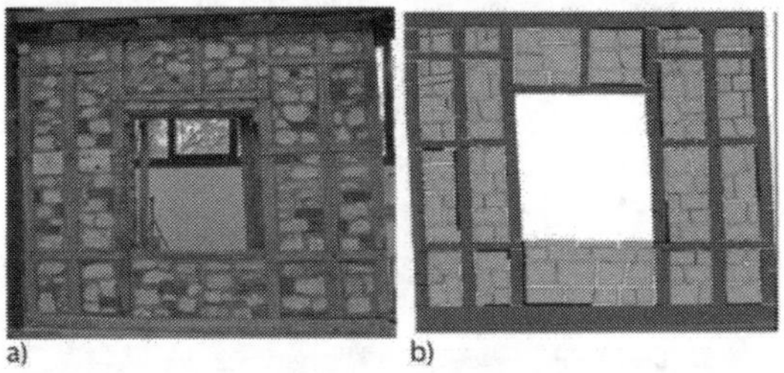
a) b)

Figure 3. Ultimate deformed shape and cracking pattern. Comparison between experimental test (a) and numerical modeling through the software *BrickWORK* (b) [8].

By means of both experimental test carried out at the CNR Ivalsa in Trento (Italy) [8] and numerical models developed through the software *BrickWORK* [9-19], the efficiency of the Bourbon building system, in its various forms and configurations, and its great adaptability to the demand for structural resilience that is required to an anti-seismic constructive element, was demonstrated, in particular regarding its response to the in-plane horizontal actions (Fig. 3). This type of structure is able to dissipate energy by means of interface frictions that are generated by the slip of stones both between the infill of masonry and the timber frame, and also thanks to some fissures generated in the mortar, as well as the expulsion of few stones. The overall timber skeleton, both elements and joints, acts, during the cycles, in elastic field. The model, devoid of the infill masonry frame, highlights a weak behaviour characterized by a high deformability under cyclic actions.

2 Palazzo Branconio. An experience of seismic retrofit

Palazzo Branconio is a very important historical building located in the historic center of the city of L'Aquila and is the result of the aggregation of a series of buildings from the Middle Ages. The shape of the current building is the result of the transformations by Girolamo Branconio in the first half of the seventeenth century (Fig. 4).

Figure 4. Palazzo Branconio, L'Aquila, Italy (picture taken from the author)

The Palace was already severely damaged by the earthquakes of 1703 and 1915, before suffering further damage from the very serious event of 6 April 2009. The spatial configuration of the building, although showing a sufficient regularity in elevation, although characterized by a decrease in thickness of the walls towards the upper floors, instead shows a substantial planimetric irregularity. This is due to the presence of walls built in different historical periods, according to distinct units and consisting of aggregates of main parts centralized on the two inner courtyards, together with a high reduction of wall thicknesses: these characteristics are the main reason of the greater concentration of crack patterns and damage in some rooms at the first floor (Fig. 5). The presence of such a heterogeneity was most likely to have provided the incipience, at the time of the earthquake, of kinematic movements caused by twisting actions, acting on the building due to the marked eccentricity between the rigid center of mass and that of the masses. The load-bearing walls of Palazzo Branconio is characterized by a typically L'Aquila brickwork technique, characterized by irregular stone elements, slightly rough-hewn, irregularly laid out, without transversal connections, with large amounts of mortar, having poor mechanical properties with fairly significant thicknesses, up to the first floor, at least 70 cm. Only in some areas of the building it is possible to find sporadic squared stone cantons.

Some on-site experimental tests on masonry samples, aimed at evaluating the mechanical properties of the material, were carried out on some walls devoid of cracks, chosen inside the building. In particular, the panels were subjected to a diagonal compression test. In Tab.1 some significant average values are reported.

Table 1. Results of diagonal compression test

Test	Cross-section thickness (cm)	τ_0 (MPa)	G (MPa)
DC – 01 Br.	70	0.033	111

The experimental results on the walls of Palazzo Branconio provided both for the shear strength τ_0 and shear elastic modulus G, values much lower than those deducible from the reference tables of the masonry technical rules, with a corresponding description of the typology of the masonry equipment (from masonry technical rules: $0.035 \leq \tau_0 \leq 0.051$; $230 \leq G \leq 480$). In such a case these masonry walls show that the typical seismic failure mode is the diagonal cracking mode, when the principal tensile stresses develop in the masonry under a combination of vertical and horizontal loads exceeding the tensile strength of the masonry materials. Obviously, for the actions out of plan the propensity of the walls described above is to offer a strong vulnerability to the presence of combined compressive and bending stress, the latter even more dangerous for the overall stability of the building.

3 Natural basalt fiber strips system (BFRP)

On the efficacy and potential for use of innovative materials such as FRP or UHTSS, high strength stainless steels, applied in the form of a grid (Reticulatus System)

a) b)

Figure 5. Front elevation of Palazzo Branconio: a) picture taken from the author; b) cracking pattern

in the field of seismic retrofit of masonry walls belonging to historical buildings, it has been widely written in the specific scientific literature [20-27].

The traditional timber frame reinforcements can be effectively replaced by a homologous frame, according to a similar geometric pattern, composed of natural basalt fiber strips, arranged in adherence to the wall front and glued to the wall itself by means of a GeoCalce (a KeraKoll patent), a hydraulic lime matrix compatible with the traditional binder present in the walls (Fig. 6). A more effective connection to the wall is obtained through the application of transversal elements, with function of diatones, equipped with gripping bow. The retrofit scheme described above (Fig. 7), eventually accompanied by a continuous basalt fiber net, placed on the opposite side of the wall, can guarantee a structural performance of exceptional efficiency in seismic conditions, without weighing down the original structure or substantially modifying the overall stiffness.

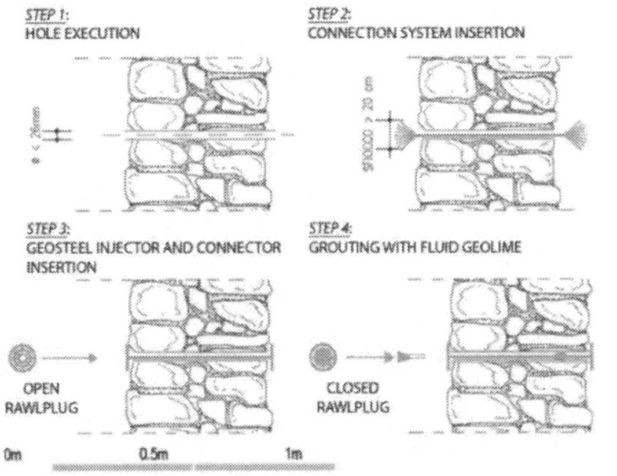

Figure 6. Application technique of natural basalt fiber strips (KeraKoll)

5 Conclusions

On the basis of examples from the past, such as the "Gaiola Pombalina" or the Bourbon concept system, re-proposing its geometric patterns and, partly, the aims of the structural performance of the elements integrated into the wall, the paper deals with the design approach for post-seismic interventions aimed to the safeguarding of historical buildings in seismic areas. The traditional timber frame reinforcements are, in this case, substituted by the natural basalt fiber strips, arranged according to similar geometric patterns, connected to the masonry walls trough transversal elements, with function of diatones, equipped by gripping bow. The system seems to be able to guarantee an excellent resistance both to the shearing actions and to the combined compressive and bending ones.

Acknowledgments

The author wishes to thank DiaCon Srl, an academic Spin-Off of the University of Florence, which has experimented, from the point of view of executive design, the technology described in this paper.

References

1. P. Vitruvio, Dieci Libri di Architettura, Libro II Cap. X, T. Carunchio (Ed.), Bardi, Roma (2006)
2. N. Ruggieri, S. Galassi, G. Tempesta, Pompeii's Stabian Baths. Mechanical behaviour assessment of selected masonry structures during the 1st century seismic events, Int. J. Archit. Herit. **12**, 5, 859-878 (2018) 10.1080/15583058.2017.1422571
3. N. Ruggieri, Timber "Science" in Herculaneum (Italy) during the 1st century A.D., Materials Science Forum (ISSN: 0255-5476), Trans Tech Publications, Proc. of 4th International Conference on Mechanical Structures and Smart Materials (4th ICMSSM2018) (22-23 September 2018, Shenzhen, China)
4. A. Amoroso, M. Bianchini, F. Di Gennaro, F. Fraioli, M. Merlo, Strutture semipogee nell'ager Fidenatis, in: Suburbium, 2. Il suburbio di Roma dalla fine dell'età monarchica alla nascita del sistema delle ville (V-II secolo a.c.), Roma 2009, pp. 347-367
5. R. Langenbach, Learning from the past to protect the future: Armature crosswalls, Eng. Struct. **30**, 8 2096-2100 (2008) 10.1016/j.engstruct.2008.04.028
6. R. Langenbach, Crosswalls instead of shearwalls: a proposed research project for the retrofit of vulnerable concrete buildings in earthquake areas based on traditional himis construction, Proc. of the Turkish 5th National Conference on Earthquake Engineering (26-30 May 2003, Istambul, Turkey)
7. E. Poletti, G. Vasconcelos, Seismic performance of traditional half-timbered walls: experimental results, pp. 53-66, in: Ruggieri N., Tampone G., Zinno R. (Eds), *Historical Hearthquake-Resistant Timber Frames in the Mediterranean Area*, Springer, Cham, 2015, pp. 95-103 (2013) 10.1007/978-3-319-16187-7
8. S. Galassi, N. Ruggieri, G. Tempesta, Seismic

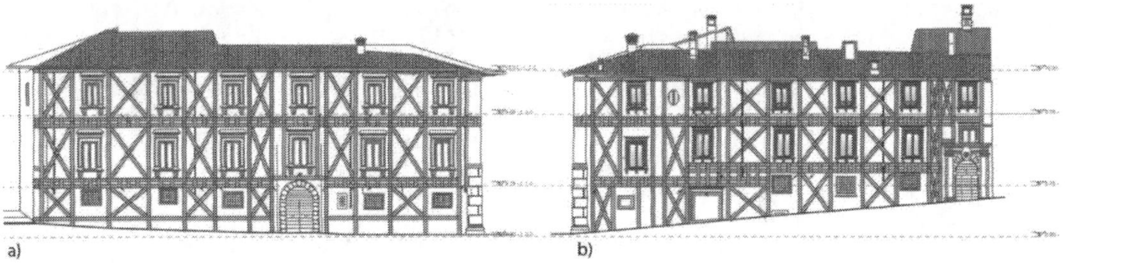

Figure 7. BFRP strengthening intervention of Palazzo Branconio : a) front elevation; b) side elevation

performance evaluation of timber-framed masonry walls. Experimental tests and numerical modelling, in: Ruggieri N., Tampone G., Zinno R. (Eds), *Historical Hearthquake-Resistant Timber Frames in the Mediterranean Area*, Springer, Cham, 2015, pp. 95-103, Proc. of 1st International Symposium on Historic Earthquake-Resistant Timber Frames in the Mediterranean Area, Cosenza, Italy (4-5 November 2013) 10.1007/978-3-319-16187-7_8

9. S. Galassi, N. Ruggieri, G. Tempesta, A novel numerical tool for seismic vulnerability analysis of ruins in archaeological sites, Int. J. Archit. Herit. (2018), 10.1080/15583058.2018.1492647
10. S. Galassi, N. Ruggieri, G. Tempesta, Ruins and archaeological artifacts: vulnerabilities analysis for their conservation through the original computer program BrickWORK, in: Aguilar, R., Torrealva, D., Moreira, S., Pando, M., Ramos, L.F. (Eds.), *Structural Analysis of Historical Constructions*, RILEM bookseries 18, Springer International Publishing, pp. 1839-1848 (2018) 10.1007/978-3-319-99441-3_197, Proc. of 11th International Conference on structural analysis of historical constructions (SAHC2018)(11-13 September 2018, Cusco, Perù).
11. S. Galassi, M. Paradiso, BrickWORK Software-Aided Analysis of Masonry Structures, IERI Procedia **7**, 62-70 (2014), Special Issue 2013 International Conference on Applied Computing, Computer Science, and Computer Engineering (ICACC 2013), USA, December 2013, Elsevier B.V., ISSN: 2212-6678, 10.1016/j.ieri.2014.08.011
12. S. Galassi, G. Misseri, L. Rovero, G. Tempesta, Equilibrium analysis of masonry domes. On the analytical interpretation of the Eddy-Lévy graphical method, Int. J. Archit. Herit. **11**, 8, 1195-1211 (2017) 10.1080/15583058.2017.1372823
13. S. Galassi, G. Tempesta, Analysis of Masonry Block Structures with Unilateral Frictional Joints, in: *Proc. Int. Conf. on Electronic, Control, Automation and Mechanical Engineering (ECAME 2017)*, 19-20 November 2017, Sanya, China, DEStech Transactions on Engineering and Technology Research, 246-51, 10.12783/dtetr/ecame2017/18396
14. F. Pugi, S. Galassi, 2013 Seismic analysis of masonry voussoir arches according to the Italian building code, Ing. Sismica-Ital. **30** ,3, 33-55 (2013)
15. S. Galassi, G. Misseri, L. Rovero, G. Tempesta, Failure modes prediction of masonry voussoir arches on moving supports, Eng. Struct. **173**, 706-717 (2018) 10.1016/j.engstruct.2018.07.015
16. M. Paradiso, S. Galassi, S. Benedetti, Stabilità di archi e volte in muratura: metodi tradizionali e calcolo automatico, in: Luis Villegas, Ignacio lombillo, Clara Liano, Haydee Blanco (Eds.), *Patologia della Costruccion, Tecnologia de la Rehabilitacion y Gestion del Patrimonio*, Proc. Conf Latinoamericano Patologie del costruito, tecniche di riabilitazione e gestione del patrimonio (Rehabend 2014), Santander, Spain, Gráficas Iguña, S.A. (2014)
17. M. Paradiso, S. Galassi, S. Benedetti, Il caratteristico Forte di San Fernando de Bocachica, Colombia, in: Luis Villegas, Ignacio lombillo, Clara Liano, Haydee Blanco (Eds.), *Patologia della Costruccion, Tecnologia de la Rehabilitacion y Gestion del Patrimonio*, Proc. Conf. Latinoamericano Patologie del costruito, tecniche di riabilitazione e gestione del patrimonio (Rehabend 2014), Santander, Spain, Gráficas Iguña, S.A. (2014)
18. S. Galassi, L. Dipasquale, N. Ruggieri, G. Tempesta, Andalusian timber roof structure in Chefchaouen, Northern Morocco: construction technique and structural behavior, J. Archit. Eng. **24**, 3, 04018012 (2018) 10.1061/(ASCE)AE.1943-5568.0000315
19. S. Galassi, N. Ruggieri, L. Dipasquale, G. Tempesta, Assessment of the Moroccan Vernacular Timber Roof: a Proposal for an Eco-Friendly Strengthening System, J. Architectural Conservation (in press).
20. M. Paradiso, S. Galassi, A. Borri, D. Sinicropi, "Reticolatus": an innovative reinforcement for irregular masonry. A numeric model, in: Paulo J. S. Cruz (Ed.), *Structures and Architecture: Concepts, Applications and Challenges*, CRC Press / Balkema, Taylor and Francis Group, London, Proc. 2nd Int. Conf. on Structure & Architecture ICSA2013, Guimaraes, Portugal (2013)
21. M. Corradi, A. Borri, G. Castori, R. Sisti, The Reticulatus method for shear strengthening of fair-faced masonry, Bulletin of Earthquake Engineering **14**, 12, 3547–3571 (2016) 10.1007/s10518-016-0006-5
22. D. Sinicropi, E. Perria, S. Galassi, M. Paradiso, A. Borri, Artificial Ageing of Mortar Prisms Reinforced through Steel, Glass and Organic Fibers, Journal of Key Eng. Mat., A. Di Tommaso, C. Gentilini and G. Castellazzi (Eds.) **624**, 4, 542-50 (2014) 10.4028/www.scientific.net/KEM.624.542
23. S. Galassi, A numerical procedure for failure mode detection of masonry arches reinforced with fiber reinforced polymeric materials, IOP Conf. Ser.: Mater. Sci. Eng. **369**, 012038 (2018) 10.1088/1757-899X/369/1/012038, Proc. of 5th Global Conference on Polymer and Composite Materials (PCM 2018) (10-13 April 2018, Kitakyushu City, Japan)
24. S. Galassi, Analysis of masonry arches reinforced with FRP sheets: experimental results and numerical evaluations, MATEC Web of Conferences 207: 01002, 10.1051/matecconf/201820701002, Proc. of International Conference on Metal Material Processes and Manufacturing (ICMMPM2018) (19-20 July 2018, Jeju Island, South Korea)
25. A. Borri, G. Castori, M. Corradi, Intrados strengthening of brick masonry arches with composite materials, Compos. Part B-Eng. 42, 5, 1164-1172 (2011) 10.1016/j.compositesb.2011.03.005
26. G. Castori, A. Borri, M. Corradi, Behavior of thin masonry arches repaired using composite materials, Compos. Part B-Eng. **87,** 311-321 (2016) 10.1016/j.compositesb.2015.09.008
27. A. Borri, G. Castori, M. Corradi, Strengthening of thin masonry arches, Key Eng. Mat. **624**, 51–58 (2015) 10.4028/www.scientific.net/KEM.624.51

Simultaneous Determination of Three Effective Components in Different Compatibility Proportion of the Couple of Sophora flavescens-Fructus Ligustri lucidi by HPLC

Shuhan Chu[1], *Hongxu* Zhang [1], *Lixin* Ding [1] *

[1] Key Lab of Biological Medical Preparation, Pharmaceutical School of Jiamusi University, Jiamusi, Heilongjiang, China, 154007

Abstract: Sophora flavescens - Ligustrum lucidum has the effect of increasing the anti-cancer effect of white blood cells and treating liver fibrosis. In order to determine the optimal compatibility ratio of this drug pair, the rationality of the drug pairs in different proportions was revealed. In this experiment, three active ingredients of the Sophora flavescens-Fructus Ligustri lucidi were extracted by reflux extraction, which were matrine, oxymatrine and specnuezhenide. A method for simultaneously determining the content of matrine, oxymatrine and specnuezhenide in a short time by HPLC was established. The method is simple, precise, and stable. Reproducibility can be used to determine the content of three active ingredients of matrine, oxymatrineand specnuezhenide. The analysis results was indicated that the different proportions of compatibility and different extraction solvents have significant effects on the content of active ingredients and dissolution rate of Sophora flavescens- Fructus Ligustri lucidi, after the combination of Sophora flavescens- Fructus Ligustri lucidi in different proportions, the content of matrine, oxymatrine and specnuezhenide is greater than the content of the active ingredients of each single herb. When the ratio of compatibility of Sophora flavescens-Ligustrum Ligustri lucidi is <1:1, it is beneficial to the dissolution of the three active ingredients. By determining the optimal compatibility ratio of this drug pair, provided a theoretical basis for the study of late pharmacodynamics and provide theoretical basis for clinical application of traditional Chinese medicine.

1 Introduction

The drug pair, is a combination of relatively fixed 2 medicinal tastes commonly used in traditional Chinese medicine, and is a basic form of compatibility of traditional Chinese medicines. Through the combination of the two drugs, synergistic effect or attenuation is played [1]. Traditional Chinese medicines will undergo complex chemical composition changes during the compatibility process. In the process of compatibility, attention should be paid to the changes in chemical composition and compatibility pharmacokinetics of each drug. Therefore, this study mainly studied the content of three active ingredients of matrine, oxymatrine and specnuezhenide under the different ratios of sophora flavescens and Fructus Ligustri lucidi, and determined the optimal compatibility ratio.

Sephora flavescens-Fructus Ligustri lucidi was taken from the "Shi Jinmo medicine", The prescription consist of Sephora flavescensa 6 ~ 10 g and Fructus Ligustri lucidi 10 ~ 15 g. Sophora flavescens has the disadvantage of immunosuppression; Fructus Ligustri lucidi has the advantage of enhancing immunity. The combination of the two drugs can increase the anti-cancer effect of white blood cells. Sophora flavescens Ait, the dry root of Sophora Flavescens Ait, has the effects of dampness, and diuresis. The main active ingredients of Sophora flavescens are matrine and oxymatrine, which have anti-tumor, anti-liver cancer, elevated leukocyte and antibacterial activity. Fructus Ligustri lucidi, the ripe mature fruit of Fructus Ligustri lucidi Ait, is nutritious for liver, kidney, and eyeshades. Specnuezhenide, salidroside, oleanolic acid were the main active ingredients of fructus igustri lucidimain active ingredients for specnuezhenide, salidroside, oleanolic acid, it has the pharmacological effects of protecting liver, anti-cancer and anti-inflammator.

The 2015 edition of the Chinese Pharmacopoeia records that matrine is based on matrine and oxymatrine as quality control standards, and Fructus igustri lucidi is a quality control standard [2]. According to the literature, the current research on Fructus igustri lucidi focuses on active components such as oleanolic acid and salidroside [3-4]. The chemical composition and pharmacodynamics of the two single-flavored drugs were studied separately. There is no report on the content determination of the different compatibility ratios of Sophora flavescens-Fructus igustri lucidi[5-7]. The HPLC method is used to determine the content of Sophora flavescens in the 2015 edition of the Chinese Pharmacopoeia, the column is an

* Corresponding author: dlx200475@163.com

https://doi.org/10.1051/matecconf/201824201015

NH_2 column, most of the literature reports that the content of matrine and oxymatrine is determined by HPLC method using a column as an NH_2 column [8-9]. Although the HPLC method is used to determine the content of Fructus igustri lucidi , but the column is a C18 column. The detection method of Sophora flavescens and Fructus igustri lucidi in the Chinese Pharmacopoeia cannot be detected at the same time. This study established a HPLC method for the simultaneous determination of three effective components in matrine, oxymatrine and specnuezhenide by HPLC with C18 column. The effects of different proportions of compatibility on the content of these three active ingredients in the drug pair of Sophora flavescens-Fructus Ligustri lucidi were compared. The dissolution rate of the active ingredient in different extraction solvents is studied, and the optimal extraction solvent of the drug pair is preferred. According to the literature, the optimal ethanol extraction concentration of single-drug is 60%-70% [10-12]. Therefore, in this experiment, water and 60% ethanol were selected as two different extraction solvents to study the effect of the active ingredient on the dissolution rate of the active ingredient of Sophora flavescens and Fructus Ligustri lucidi.

2 Materials and methods

2.1Chemicals, reagents and materials

Matrine reference product (batch number: N0622AS, purity >99.8%), oxymatrine reference product (batch number: S0715AS, purity >99.8%), specnuezhenide reference product (batch number: M0621AS, purity >99.8%). They were all purchased from Dalian Meilun Biotechnology Co., Ltd.; acetonitrile and methanol were chromatographically pure, The remaining reagents are of analytical grade, and water was purified.

Model 1100 HPLC (Agilent Corporation, USA), DS-T100 Traditional Chinese Medicine Crusher (Shanghai dingshuai instrument co. LTD), CP2250 Electronic Analytical Balance (Shanghai orhaus international trading co. LTD), Sophora flavescens and Fructus Ligustri lucidi were purchased from Jiamusi City Jintian pharmacy.

Sophora flavescens and Fructus Ligustri lucidi were purchased from Jintian Pharmacy of Jiamusi City. They were identified by Zong Ximing, senior pharmacist of Pharmacy Experimental Center of Jiamusi University School of Pharmacy, as dry roots of the leguminous plant, and dry fruits of the oleaceae plant.

2.2 Preparation of mixed reference solution

The amount of matrine, oxymatrine, and specnuezhenide was accurately placed in a 5 mL volumetric flask; Methanol was added to volume to obtain a mixed standard solution with concentrations of 50 g/mL, 0.15 mg/mL, and 0.25 mg/mL.

2.3 Preparation of test solution

According to the ratio of Sophora flavescens to Fructus Ligustri lucidi (m/m) of 1:1, 1:2, 2:1, 1:3, 3:1, 1:0 and 0:1, The total amount of the drug pair is 12g, weigh two copies of each compatibility ratio, were placed in a 250 ml round bottom flask, the amount of distilled water and 60% ethanol solution was added 4 times, reflux extraction was Performed twice for 3 h each time, the extracts were combined and filtered. Methanol were used to concentrate the filtrate and dissolved in a 50 ml volumetric flask, then drawn in a 10.0 ml and 25 ml volumetric flask, adding methanol to a constant volume, where the solution was Penetrate 0.22 m micro porous membrane and the filtrate was taken, which was the test solution.

2.4 HPLC system and conditions

The HPLC system used was an Agilent Technology 1100 series HPLC model equipped with G1311A quaternary pump, G1379A vacuum inline degasser, G1315DAD detector, G1316A column thermostat (USA). And the column was an Agilent 20RBAX XDB-C18 (250 nm × 4.6 nm). The mobile phase was acetonitrile (A)-0.2% aqueous solution of phosphoric acid (B), gradient elution (0 to 15 min, 7% A; 15 to 20 min, 7% A→20% A; 20 to 40 min, 20% A) The flow rate was 1.0 mL·min-1; the detection wavelength was 220 nm; the column temperature was 28°C; the injection volume was 10ml. Under the above chromatographic conditions, the chromatograms of the mixed standard solution and the test sample are shown in Figure 1.

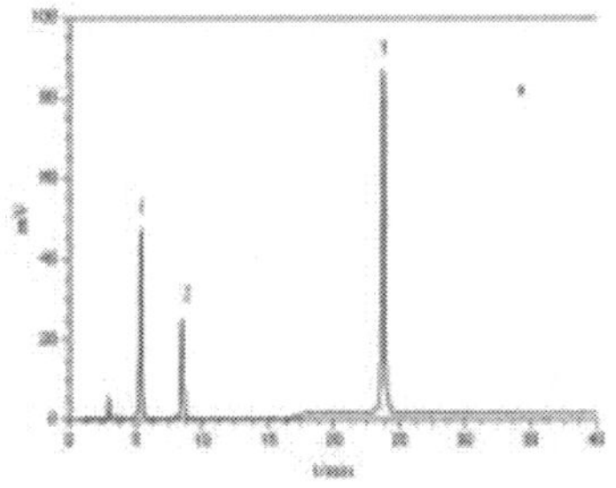

a. Mixed control: 1.matrine; 2. oxymatrine 3. specnuezhenide;

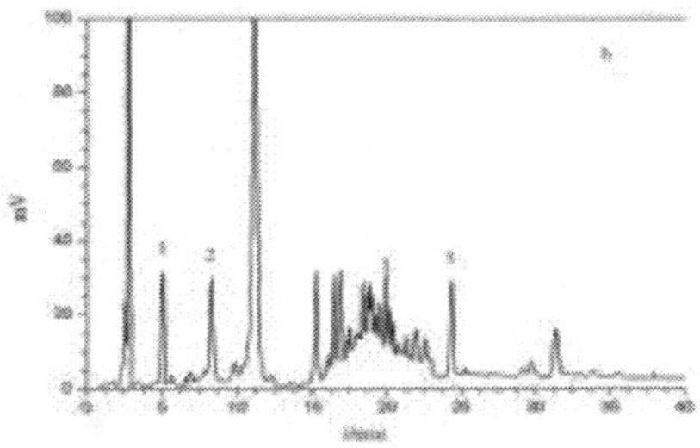

b. Test sample

Fig. 1. HPLC chromatograms of Standards and samples.

2.5 Statistical analysis

Statistical analysis was performed using SPSS 17.0 statistical analysis software. The measurement data was expressed as X±S, using one-way variance. The comparison between groups and analysis of variance was performed. The orderly qualitative data were compared by groups using nonparametric test methods. The difference was statistically significant with $P < 0.05$.

3 Results and discussion

3.1 Optimization of chromatographic conditions

Mobile phase selection: The effects of methanol-water solution and acetonitrile-water solution on the analysis results were investigated, the analysis results indicated, when methanol-water solution was used as the mobile phase, Retention time of each component was delayed, and the separation Can't satisfied Quantitative requirements. When acetonitrile-aqueous solution was used as the mobile phase, although the peaks of each component can be completely separated, however, there was tailing phenomenon, therefore, the modifier phosphoric acid was added to the aqueous solution. Further examination of the amount of phosphoric acid, by comparison, when acetonitrile-0.2% phosphoric acid aqueous solution was used as the mobile phase, the peaks of the component to be measured can be completely separated, the retention time is early, and the shape is better. In order to accurately determine the content of the three active ingredients of matrine, oxymatrine, and specnuezhenide, gradient elution was used.

Detection wavelength selection: Establishment of HPLC method to examine the detection wavelength of each component, the maximum absorption wavelength of matrine and oxymatrine in the 2015 edition of the Chinese Pharmacopoeia is 220 nm. The maximum absorption wavelength of the specnuezhenide was 224 nm. Another report in the literature that specnuezhenide have a greater absorption at wavelengths of 220 nm [13]. Through pre-testing, it was proved that the specnuezhenide area was basically the same at the wavelengths of 220 nm and 224 nm. In order to detect the contents of three active ingredients of matrine, oxymatrine, and specnuezhenide at the same time more quickly and conveniently, 220 nm was selected as the detection wavelength in this experiment.

3.2 Linearity and range

Precise measurement of the reference solution under "2.2" 1.0, 2.0, 3.0, 4.0, 5.0 ml, placed in a 5 mL volumetric flask. Then methanol was added to volume of a series of mixed reference solutions. Accurately 1.0 l of each of the above mixed control solutions was measured to determine the chromatographic conditions under "2.4" and the peak area was recorded. With the mass concentration of the component to be measured (g/ml) as the abscissa, a standard curve was drawn, the regression equations for matrine, oxymatrine, and specnuezhenide were $y=12.392x-3.7452(r=0.9971)$, $y=33.89x-3.975$ $(r=0.9983)$, and $y=9.4213x-4.3388(r=0.9994)$.

Table 1. Determination of alcohol extract sample content (x ± s, n = 3)

A:B	Matrine Content (mg/g)	Matrine dissolution rate (%)	Oxymatrine content (mg/g)	Oxymatrine dissolution rate (%)	Specnuezhenide content (mg/g)	Specnuezhenide dissolution rate (%)
3:1	0.1017±0.012	2.259	0.1734±0.021	3.853	0.1005±0.005	6.699
2:1	0.0876±0.007	2.191	0.1485±0.013	3.711	0.1405±0.009	7.027
1:1	0.0629±0.004	2.095	0.1066±0.004	3.553	0.2339±0.001	7.798
1:2	0.0400±0.014	1.999	0.0796±0.023	3.978	0.3650±0.004	9.125
1:3	0.0282±0.009	1.883	0.0566±0.018	3.776	0.3959±0.021	8.798
1:0	0.0096±0.002	0.160	0.0309±0.018	0.515	0	0
0:1	0	0	0	0	0.0660±0.019	1.099

Table 2. Determination of water extract sample content (x ± s, n = 3)

A:B	Matrine Content (mg/g)	Matrine dissolution rate (%)	Oxymatrine content (mg/g)	Oxymatrine dissolution rate (%)	Specnuezhenide content (mg/g)	Specnuezheni dedissolution rate (%)
3:1	0.1347±0.006	2.994	0.4927±0.013	10.95	0.1156±0.009	7.703
2:1	0.1134±0.012	2.834	0.4041±0.011	10.10	0.2228±0.012	11.14
1:1	0.0843±0.019	2.810	0.2905±0.028	9.683	0.4154±0.017	13.85
1:2	0.0482±0.001	2.408	0.2514±0.008	12.57	0.5761±0.013	14.40
1:3	0.0387±0.020	0.861	0.1704±0.012	11.36	0.5822±0.009	12.94
1:0	0.0096±0.017	0.160	0.0442±0.027	0.737	0	0
0:1	0	0	0	0	0.0801±0.025	1.335

3.3 Precision test

1.0 ml of the mixed reference solution under"2.2"was taken, and injected 6 consecutive times according to the chromatographic conditions under"2.4"to record the peak area. As a result, the RSDs of the peak areas of matrine, oxymatrine, and specnuezhenide were 1.1%, 1.2%, and 1.1% (n=6), respectively, and the results showed that the precision of the apparatus was good.

3.4 Stability test

The same test substance solution under"2.2"Sophora flavescens-Fructus ligustri lucidi (1:3, m/m) was taken and placed at room temperature for 0, 2, 4, 8, 16, 24 h, then according to the "2.4" under the chromatographic conditions for measurement, the peak area was recorded. As a result, the RSDs of the peak areas of matrine, oxymatrine, and specnuezhenide respectively were 1.0%, 1.2% and 1.1% (n=6), it shows that the test solution has good stability at room temperature for 24 hours under the conditions of this test.

3.5 Repeatability test

Accurately the appropriate amount of Sophora flavescens - Ligustrum lucidum (1:3, m/m) were weighed and test solution was prepared according to the method under "2.3". 6 Parallel copies made to determine the chromatographic conditions under "2.4" and the peak area was recorded. As a result, the RSDs of the peak areas of matrine, oxymatrine, and specnuezhenide were 0.95%, 1.2%, and 1.1% (n=6), indicating that the method has good reproducibility.

3.6 Sample recovery test

An appropriate amount of the known content of Sophora flavescens -Fructus Ligustri licidi (1:3, m/m) was taken. A total of nine and three groups were made. Mixed reference products were added respectively from low, medium and high quality. Prepared the alcohol solution according to the method of "2.3" to provide the sample solution, according to the chromatographic conditions under "2.4", the peak area was recorded. As a result, the range of the recovery rate of the matrine, oxymatrine, and specnuezhenide areas were 95.36%, 96.43% and 95.43%. It shows that the recovery rate of the method is good and feasible.

3.7 Determination of sample content

Appropriate amount of Sophorae flavescens - Ligustrum lucidum were taken and the test solution was prepared according to the method under "2.3", then according to "2.4" under the chromatographic conditions for determination, the peak area was recorded , the content and dissolution of the sample was calculated, the results are shown in Table 1 and Table2. A is Sophora flavescens, B is fructus ligustri lucidi.

3.8 Analysis of results

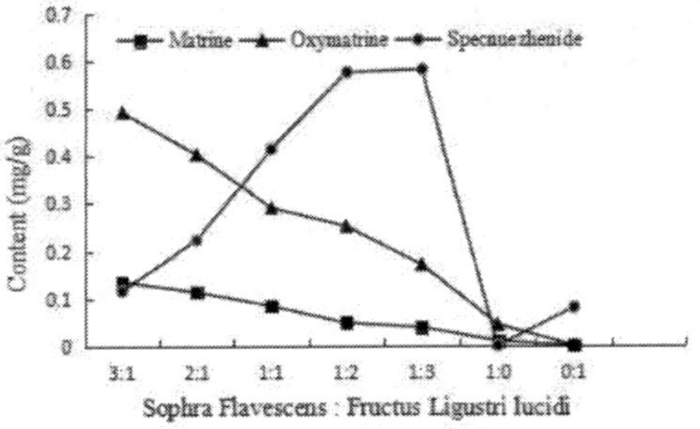

Fig.2. Contents of active constituents in different compatibility ratios of Sophora flavescens and Ligustrum lucidum during ethanol extraction.

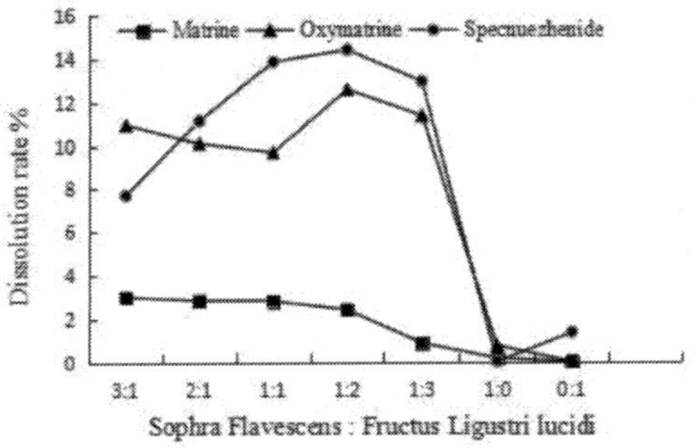

Fig.3. Dissolution rate of active ingredients in different ratios of Sophora flavescens and Ligustrum lucidum during ethanol extraction.

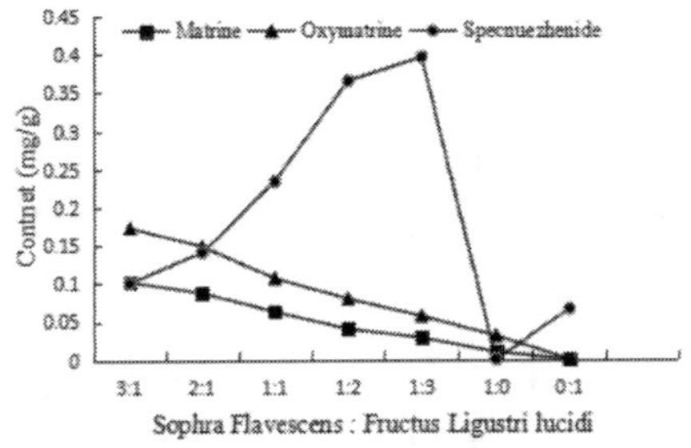

Fig.4. Contents of active constituents in different compatibility ratios of Sophora flavescens and Ligustrum lucidum during water extraction.

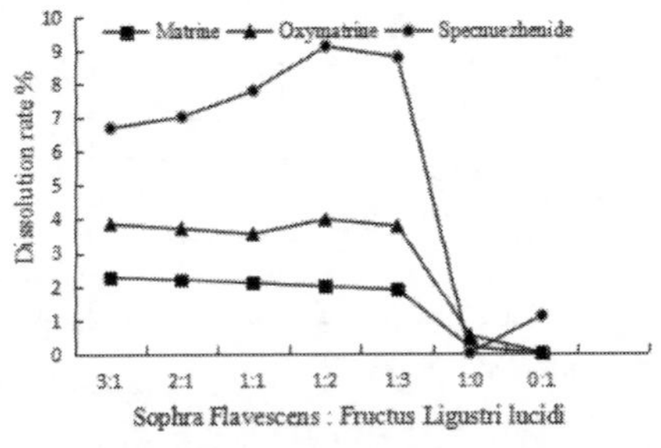

Fig.5. Dissolution rate of active ingredients in different ratios of Sophora flavescens and Ligustrum lucidum during water extraction.

Table 1 shows Figure 2 and Figure 3, Table 2 shows Figure 4 and Figure 5. It can be concluded that the active ingredient in the extraction of Sophora flavescens-Fructus Ligustri lucidi is superior to the water extraction method by using 60% ethanol. When water extracting method was used to extract Sophora flavescens-Fructus Ligustri lucidi, had precipitate formation. When 60% ethanol was used to extracted, the precipitation was reduced. It may be caused by that 60% ethanol solvent promotes the dissolution of the precipitate and increases the content of the active ingredient. The content and dissolution rate of the active ingredients in the different ratios of Sophora flavescens-Fructus Ligustri lucidi were significantly greater than the active ingredients and dissolution rate of the single-flavored medicines, which increased with the increase of the single-flavored pharmaceutical dosage. The content of matrine and the dissolution rate are not significantly different under different ratios. The effect of the active ingredients oxymatrine and specnuezhenide is significant. When the compatibility ratio is 1:2, the dissolution rate of oxymatrine and specnuezhenide is the largest. Xie Ying [14] and other believe that the compatibility is not a simple 1+1 addition, but in the process of boiling, its chemical properties have undergone qualitative and quantitative changes, thus exerting the unique effects of the compound. Therefore, sophora flavescens and fructus ligustri lucidi may undergo some chemical reaction after compatibility, which promotes the dissolution of matrine, oxymatrine and specnuezhenide, and increases its content. "Shi Jinmo medicine" records 6-10g of Sophora flavescens and 10-15g of Fructus Ligustri lucidi. It can be seen that the dosage of Sophora flavescens is lower than that of Fructus Ligustri lucidi, so the compatibility ratio of 1:2 is in line with the scope of the drug. The effects of different compatibility ratios on the content of other ingredients in the Sophora flavescens- Fructus Ligustri lucidi and the pharmacodynamics study need to be further explored to maximize the efficacy.

4 Conclusions

In conclusion, HPLC methodestablished in this paper is simple, precise, stable and reproducible. It can be used for the simultaneous determination of three active ingredients in the drug pair of Sophora flavescens-Fructus Ligustri lucidi, such as matrine, oxymatrine and specnuezhenide.This method is quick and easy. Provided a theoretical basis for the study of late pharmacodynamics. The drug pair of Sophora flavescens-Fructus Ligustri lucidi on different ratios can promote the dissolution of these three active ingredients. The active ingredient in the extraction of Sophora flavescens-Fructus Ligustri lucidi is superior to the water extraction method by using 60% ethanol. When the ratio of compatibility of Sophora flavescens-Ligustrum Ligustri lucidi is <1:1, it is beneficial to the dissolution of the three active ingredients.

This work was supported by the Jiamusi University President Innovation and Entrepreneurship Fund [grand number XZYF-2018-44].

5 References

1. Y.P. Tang, X.Y. Shu, W.X. Li, M.Zhu, S.L.Xiu, D.W.Qian, X.S.Fan, J.A.Duan, Drug research (I) - the formation and development of drug pairs [J]. Chinese Journal of Traditional Chinese Medicine, **38**, 4185-4190 (2013) (in Chinese)
2. National Pharmacopoeia Commission. Pharmacopoeia of the People's Republic of China: one [s]. 2015 edition. Beijing: China Medical Science and Technology Press, 202 (2015) (in Chinese)
3. Y. Wang, Y. Chen, Q.Q.Nie. L.Ling, Y.Zhang, Extraction Process of Oleanolic Acid from Ligustrum lucidum and Its Antioxidant Activity[J]. Jiangsu Agricultural Science, **45**, 174-176 (2017) (In Chinese)
4. Z.Y.Li, Q.Li, J.Lü,J.H.Ling, X.H.Yu, X.H.Chen, K.S.Bi, LC-MS determination and pharmacokinetic study of salidroside in rat plasma after oral administration of suspensions of traditional Chinese medicine Erzhi Wan and Fructus Ligustri lucidi[J]. Journal of Pharmaceutical Analysis, **1**, 8-12 (2011) (In Chinese)
5. H. L.Seo, S.Y.Baek, E.H.Leea, J.H.Lee, S.G.lee,K.Y.Kim, M.H.Jang, M.H.Park, J.H.Kim, K.J.Kim, H.S.Lee, S.C.Ahn, J.R.Lee, S.J.Park, S.C.Kim, Y.W.Kim, Liqustri lucidi Fructus inhibits hepatic injury and functions as an antioxidant by activation of AMP-activated protein kinase in vivo and in vitro[J]. Chemico-Biological Interactions, **262**, 57-68 (2017)
6. Q.Li, Y.S.Fan, Z.Q.Gao, K.Fan, Z.J.Liu, Effect of Fructus Ligustri Lucidi on osteoblastic like cell-line MC3T3-E1[J]. Journal of Ethnopharmacology, **170**, 88-95 (2015)
7. Y. Zhu, Y.N.Yang, Z.M.Feng, J.S.Jiang, P.C.Zhang, Sophoflavanones A and B, two novel prenylated flavanones from the roots of Sophora flavescens[J]. Bioorganic Chemistry, **79**, 122-125 (2018)
8. W.L. Wang, H.X. Zhao, H.Y. Jia, Z.M.Yu, D.H.JuDetermination of matrine and oxymatrine in Sophora flavescens L. from different habitats by HPLC[J]. Chinese Journal of Basic Medicine in Traditional Chinese Medicine, **13**, 799-800 (2007) (In Chinese)
9. K. Li, S.D. Wang, L.Q. Wu, H.Tang, Simultaneous determination of matrine and oxymatrine in Radix Sophorae by HPLC [J]. Chinese Journal of Pharmaceutical Sciences, **30**, 55-57 (2014) (In Chinese)
10. D. Wang, J. LV, Z.Q. Zhang, L.Zhao, Optimization of extraction process of matrine alkaloids by

orthogonal test [J]. China Journal of Traditional Chinese Medicine and Information, **18**, 51-52(2011) (In Chinese)

11. Q. Zhang, W.J. Shi, X.H. Lian, W.J.Du, A,L,Cao, Optimization of the extraction process of Sophora flavescens [J]. Shanxi Traditional Chinese Medicine, **33**, 1408-1409 (2012) (In Chinese)
12. Y.Yang, J.Yi, L.Ma,J.G.Wu, J.Z.Wu, Y.B.Wu, Optimization of extraction process of iridoid glycosides from Ligustrum lucidum by multi-index orthogonal test [J]. Chinese Pharmacy, **27**, 3970-3973. (2016) (In Chinese)
13. J.Y. Zhang, Y. Wang, X.N. Song, Z.Y.Zhang, Simultaneous determination of salidroside, scutellarin and tetraside in Ligustrum lucidum from different producing areas in Guizhou by HPLC[J]. Guizhou Medicine, **41**, 1038-1040.(2-17) (In Chinese)
14. Y. Xie, F.Q. Tang, W.J. Xu, H.F.Zhou, J.B.Luo, Effects of different compatibility ratios on the content of active ingredients in ephedra-almonds [J]. Shi Zhen Guo Yi Guo Yao, **23**, 2686-2688 (2012) (In Chinese)

Peculiarities of the use of siliceous raw materials of the Russian Far East in the integrated pipeline protection

Elena A. Yatsenko[1], Boris M. Goltsman[1,a], Anna V. Ryabova[1] and Victoria A. Smoliy[1]

[1]*Platov South-Russian State Polytechnic University (NPI), 346428 Novocherkassk, Russia*

Abstract. Modern trends in the development of Russian oil and gas infrastructure are examined. The important role of the Far East in the transportation and export of oil is revealed. The main threats in the operation of pipelines are described. The integrated protection technology of pipeline surfaces is proposed. The structure and properties of local silicate raw material – diatomite – are studied. The technology of obtaining glass enamel coating is designed to protect the internal surface of the pipe. The phase composition, microstructure and properties of the coating are compared with analogues. The technology of foam glass production is designed to protect the external surface of the pipe. The foaming processes are studied; the properties and structure of the material are examined. The optimum ratio of raw materials is revealed. Recommendations on the application of the developed technology for integrated pipeline protection are given.

1 Introduction

Currently, the main sources of energy are oil, natural gas and coal. Oil fuel due to the convenience of its transportation is the most valuable for vehicles, which are the main oil consumers. Over the past 20 years, world energy consumption has increased by 30%, and this growth is likely to continue due to the growing demand of the rapidly developing countries of the Asian region.

The energy industry is one of the most important structural components of the Russian economy. The main center of new industrial development is the Far East [1], which performs an important transit function. There is an export of oil through its territory from other regions of the country, first of all – from Siberia. In addition, it borders with China, which is one of the largest consumers of Russian hydrocarbons.

Transportation of oil and gas is carried out through steel pipelines of various diameters and assortments. The safety of pipeline operation strongly depends on the quality of the corrosion protection. Corrosion elements represent a particular danger as a source and amplifier of a wandering direct current, which leads to a perforating destruction of the metal contacting with the electrically conductive aggressive medium. The aggressive impact on the pipe wall is caused by oil from the inside and the soil from the outside. Internal protection is designed to counteract corrosion of pipes under the influence of a transported aggressive medium, and to reduce the hydraulic resistance of pipelines and increase their throughput. External protection is designed not only to maintain the integrity of pipes, the quality of oil products and the speed of their transportation, but also to protect human health and the environment. Therefore, additional requirements are imposed to external protection: prevention of condensation and high temperatures on the surface of the pipe; elimination of noise generated by pressure drop inside the pipe; prevention of temperature losses leading to solidification of transported oil; minimal impact on the environment.

The most reliable and versatile materials for the protection of pipelines are silicate materials, that can be used for manufacturing of a wide range of materials and coatings. Silicate-enamel coatings are the most promising for internal protection, due to their high chemical, thermal, corrosion and abrasion resistance. They also do not allow deposits on the pipe walls, reliably operate at temperatures from -50 to +350 ° C and provide anticorrosion protection of pipelines for at least 50 years.

The only material eligible for the external protection is foam glass – a thermal insulating material, which is literally a foamed glass mass. Along with excellent thermal insulation properties and full ecological, fire and hygienic safety, foam glass has the following range of properties: durability, incombustibility, ease of handling and installation, stability, resistance to atmospheric and biological effects, etc. In addition, foam glass materials can be obtained on the basis of a wide range of raw materials [1-7].

It is important to note that any large-scale production requires a local raw material base. The possibility of using amorphous raw materials in both silicate enamel technology [8-9] and foam glass technology [10-13] has been confirmed by a number of studies. Therefore, the presence of extensive deposits of amorphous silica [14] and silica-bearing rocks in Eastern Siberia and the Far

[a] Corresponding author: boriuspost@gmail.com

East, as well as the possibility of synthesizing amorphous silica from quartz-containing raw materials [15] makes it possible and necessary to develop the proposed technology for integrated protection of pipelines in the evolving network of oil and gas pipelines of the Russian Far East.

2 Materials and Methods

The main criteria of raw material selection for the study were the maximum amount of silicon oxide and the minimum content of impurities. The following raw materials were chosen: the diatomite of the Chernoyarsky deposit and the quartz sand of the Chalgan deposit. All deposits are located in the Khabarovsk Territory in a small distance from the cities and the nearest passage of oil pipelines. The chemical composition of the selected raw materials is given in Table 1.

Table 1. Chemical composition of raw materials.

Material	Chemical composition, [wt. %]					
	SiO_2	Al_2O_3	Fe_2O_3	CaO	MgO	Calci-nation losses
Diatomite	76,2	6,8	3,5	1,0	0,9	11,6
Quartz sand	99,7	–	0,02	–	–	0,3

Enamel frits were vitrified in alundum crucibles in an electric furnace at a temperature of 1100-1350 °C with an exposure of 1 h [16-17]. Cooling of the melt was carried out by granulation in cold water. The frits milled to 63 µm fraction (mesh No. 230) and then used to prepare slurry suspension consisting additionally: water – 40 %, refractory clay – 4 %, electrolyte – 0,1 %. The slurry was applied to 1008 steel samples using the dipping method. The excess slurry was removed, and then the samples were dried for 15 minutes at a temperature of 75-100 °C. Then the samples were fired in an electric furnace at a temperature of 850 °C with an exposure time of 3 min.

The softening point and flowability were determined on pressed cylindrical samples of a frit. The samples were heated and sequentially extracted from the furnace after each 10 °C to determine the softening point. The temperature at which the melting (decreasing the height) of the sample begins is the softening point. To determine flowability, the samples were placed on prepared steel plates, heated up to 850 °C, held for 2 minutes, sharply tilted to an angle of 45 °, followed by holding for 2 minutes. After cooling, the length of the sample was measured.

Temperature coefficient of linear expansion (TCLE) of enamel was determined on a vertical dilatometer DKV-4 in the temperature range 50-400 °C by measuring the relative elongation of the sample upon heating (Eq. 1):

$$\alpha = (\Delta t \cdot \Delta l \cdot 10^{-3} / l_0) + \alpha_q \quad (1)$$

where α – TCLE; Δt – temperature difference, °C; Δl – sample elongation, mm; l_0 – initial sample length, mm; $\alpha_q = 5{,}5 \cdot 10^{-7}$ °C^{-1} – TCLE of quartz glass.

Production of foam glass was conducted by the standard powder method. Raw materials (glass cullet, diatomite) were pre-dried at 120 °C. The resulting dry powders were milled to 210 µm fraction (mesh No. 70). Foaming mixture was prepared in a separate vessel by mixing the components (waterglass, glycerol, water) in a given proportion. Afterwards, the prepared raw materials were composed, mixed and molded into cubes with edge length of 20 mm and mass of 10 g (volume $8 \cdot 10^{-6}$ m^3, density 1250 kg/m^3). Then samples were loaded into a furnace for heat treatment according to [18-20].

When the air inside the furnace cooled to room temperature, the samples were removed from the furnace and subjected to mechanical processing (filing) to give them a regular shape. Further, the weight of the samples of given shape was measured. Then calculations of the volume *(V, cm³)*, apparent density *(D, kg/m³)* and foaming coefficient *(FC)* were performed based on the obtained data according to Eq. 2 – Eq. 4, respectively.

$$V = a \cdot b \cdot c \quad (2)$$

$$D = m / V \cdot 1000 \quad (3)$$

$$FC = V / 8 \quad (4)$$

where *a* – sample length, cm; *b* – sample width, cm; *c* – sample height, cm; *V* – sample volume, cm^3; m – sample mass, g.

Each recorded testing value was the mean of the results from five samples.

Analysis of the sample microstructure was performed on a TEScan VEGA II LMU scanning electron microscope with INCA ENERGY 450/XT energy dispersive microanalysis (Silicon Drift (ADD) detector). Samples were placed on a conductive carbon tape; the surface is deposited with carbon. The differential thermal analysis was carried out on a STA 449 Jupiter device in platinum crucibles with a cover. Measurement mode: temperature range 30-1000 ° C; heating rate (10 °C/min); medium: air. X-ray phase analysis was performed on a diffractometer "DRON-7". Measurement mode: scanning range 10.00-90.00 °; scanning method 2THETA-THETA; exposure 1 sec; step 0,050 °; radiation of Cu (29); wavelength 1.5406 Å.

3 Results and discussion

An X-ray phase analysis was performed for a preliminary assessment of the possibility of using diatomite as raw material to produce silicate materials, shown in Figure 1.

It can be seen that the diatomite is composed of an amorphous substance with crystalline formations consisting of various modifications of SiO_2 (mostly, β-quartz) and small quantity of $(Al_2O_3)_m \cdot (SiO_2)_n$ minerals. To study the features of phase and structural transformations during the heat treatment process, a differential thermal analysis was performed, its results are shown in Figure 2.

 https://doi.org/10.1051/matecconf/201824201016

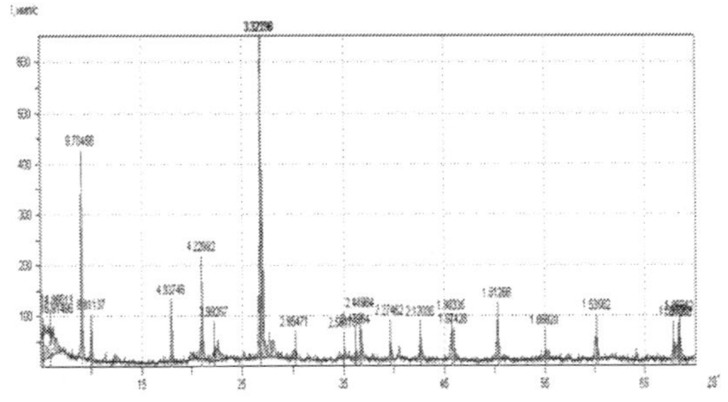

Figure 1. Results of X-ray phase analysis of diatomite.

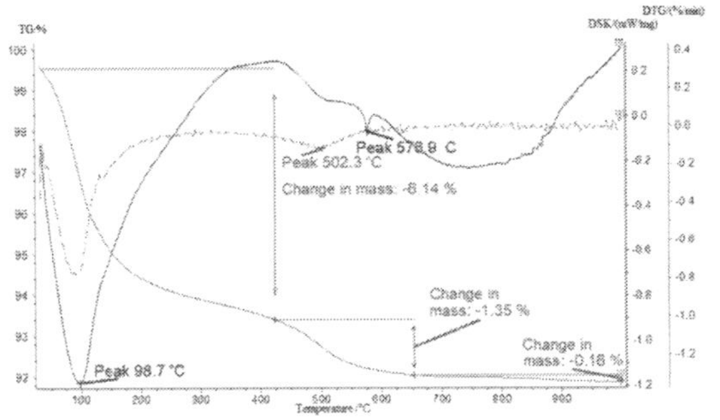

Figure 2. Results of differential thermal analysis of diatomite.

The presence of the endothermic effect in the temperature range up to 200 °C on the DTA curves indicates the removal of adsorption water, which is confirmed by indicating its smooth continuous dehydration during heating up to 1000 °C, with the main part of the water being already removed before 425 °C. The appearance of an endothermic effect on the curve with a minimum at 577 °C is due to the polymorphous transformation of crystalline quartz.

3.1. Glass enamel coating

Three glass compositions in the system «R_2O – RO – SiO_2 – Al_2O_3 – B_2O_3 –TiO_2 – CaF_2 – Na_3AlF_6 – Fe_2O_3 – Co_2O_3» were synthesized to study the diatomite suitability for the enameling of steel pipelines. In these compositions, the raw material for the introduction of silicon oxide SiO_2 was changed: in the A1 composition it was introduced by chemically pure raw material – silicon hydrate $SiO_2{\cdot}nH_2O$, in the A2 composition – by quartz sand, in A3 – by partial replacement of quartz sand with diatomite.

During melting, it was found that the processes of glass formation occur more intensively and at a lower temperature in composition A3 (synthesis mode 1100 °C, 5 hours), while compositions A1 and A2 were synthesized at 1350 °C, 7 hours. It can be explained by the fact that in compositions A1 and A2 silica is introduced by crystalline compounds – silicon hydrate and sand, which are characterized by a higher structural strength. At the same time in composition A3 most of the silica is introduced by its amorphous form. Then X-ray phase analysis of coatings was performed to study their phase composition, the results are shown in Figure3.

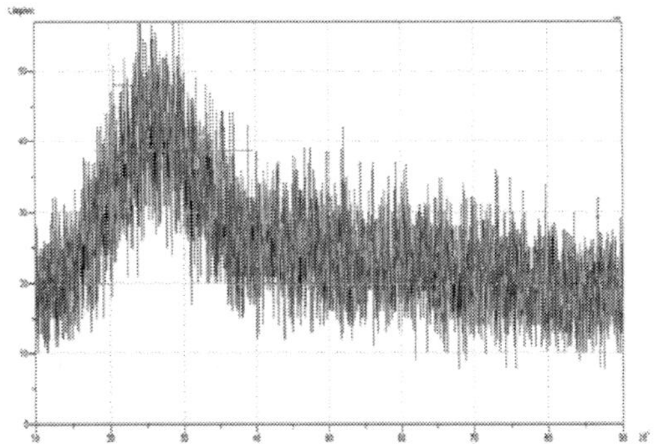

Figure 3. X-ray patterns of glass enamels: black – A1 composition, gray – A2 composition, blue – A3 composition.

Obtained results show that all coatings are in an amorphous state and do not have pronounced peaks, which indicates a homogeneous structure of the coatings. Technological properties of synthesized enamel frits (determined experimentally) are presented in Table 2.

Table 2. Technological properties of the A series samples.

Composition	TCLE at 400 °C, [$°C^{-1}{\cdot}10^{-7}$]	Softening point, [°C]	Flowability, [mm]
A1	105,07	510	47
A2	107,32	480	46
A3	103,69	450	49

The obtained results show that all synthesized frits are characterized by similar characteristics and enter the intervals permissible for silicate enamel coatings. However, a comparative analysis shows that partial replacement of silicon hydrate and quartz sand with diatomite significantly reduces melting temperature and the viscosity of glass melt. In addition, the frit obtained with amorphous silica is characterized by greater flowability, which will contribute to a better uniform formation of the enamel coating These observations are confirmed by the results of a study of the microstructure of the coatings (Figure 4).

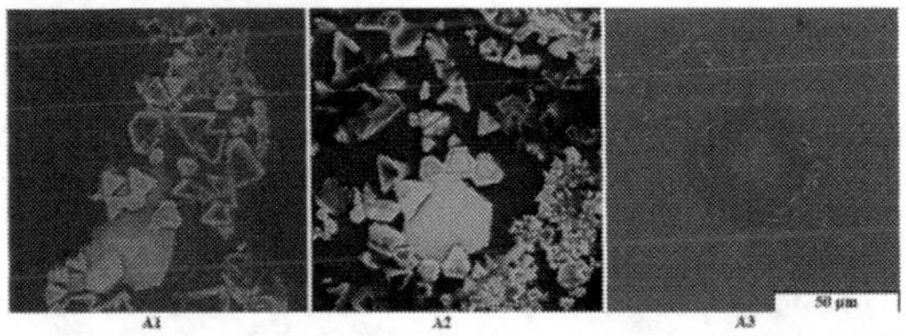

Figure 4. Microstructure of glass enamel coatings A1-A3.

The amorphous vitreous phase in the photographs is represented by dark areas, and the crystalline phase in the form of crystals of different size. It can be seen that the structure of the coating is most uniform in A3 composition based on diatomite. In samples of other compositions, crystalline inclusions of various sizes and geometric shapes were observed. Thus, partial or complete replacement of quartz sand with amorphous silica materials, in particular diatomite, reduces the duration of the silicate and glass formation by 30 % and reduces energy consumption due to the reduction of the maximum temperature by 200 °C.

3.2 Foam glass

According to the obtained results, a number of compositions of the B series were developed to study the possibility of using silicate raw materials of the Far East in the production of foam glass. The compositions include the following raw materials: colorless glass cullet, diatomite of the Chernoyarsky deposit. The ratio of materials is shown in Table 3. The pore-forming mixture was introduced in an amount of 10 wt. % over 100.

Table 3. Technological properties of the A series samples.

Component	Component content, wt.%, in composition, #				
	B1	B2	B3	B4	B5
Diatomite	20	40	60	80	100
Glass cullet	80	60	40	20	0

Samples of the B series were heat-treated at the foaming temperatures 800, 850, 900 °C. The structure of the samples obtained at 850 °C is shown in Figure 5, the properties of the samples are shown in Table 4.

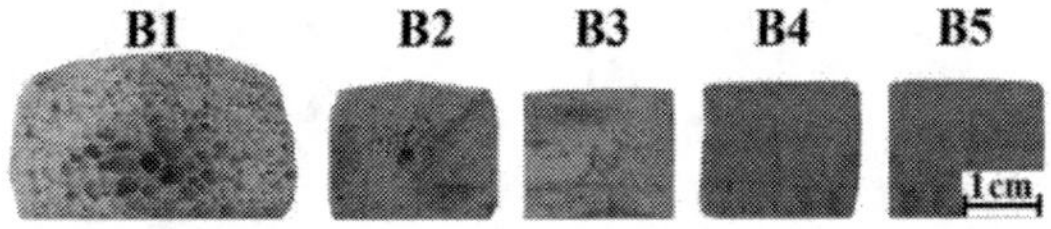

Figure 5. Internal structure of B series samples (foaming temperature 850 °C).

Table 3. Technological properties of the A series samples.

	B1	B2	B3	B4	B5
Foaming temperature 800 °C					
Density, [kg/m^3]	385,25	950,37	1401,87	952,05	941,74
Foaming coefficient	2,81	1,20	0,81	1,08	1,05
Foaming temperature 850 °C					
Density, [kg/m^3]	235,62	769,07	1633,85	1135,67	1135,09
Foaming coefficient	3,82	1,36	0,66	0,91	0,82
Foaming temperature 900 °C					
Density, [kg/m^3]	211,30	723,17	1627,29	1113,85	1243,18
Foaming coefficient	3,53	1,53	0,61	0,85	0,70

It should be noted that during the thermal processing of silicate materials, the sintering and melting processes are carried out successively. In the case of foam glass, sintering is accompanied by a decrease in volume due to the contraction of the batch particles, and melting is accompanied by an increase in volume due to foaming of the softened glass mass with gases from the foaming agent. The results obtained demonstrate that introduction of 20 wt. % of diatomite (B1) allows to obtain foam glass with a density of about 200 kg/m^3. Then foaming noticeably deteriorates with the addition of 40 wt. % of diatomite (B2), and the density increases to 700 kg/m^3. Composition B3 with 60 wt. % of diatomite has the highest density among all of the samples because of intense sintering, but its viscosity does not allow the pore formation. In compositions B4 and B5, both foaming and sintering processes are absent.

For the foaming of compositions B3-B5, it is necessary to introduce flux additives which intensifying the sintering and melting of the material. Therefore, the flux (NaF) in an amount of 10 wt. % over 100 was added in the samples of composition B3-B5 (compositions B3*-B5*, respectively). The samples were subjected to heat treatment at foaming temperatures of 800, 850, 900 ° C. Structure of the obtained samples are shown in Fig. 6, comparison of the phase composition of samples B1 and B5* - in Figure 7.

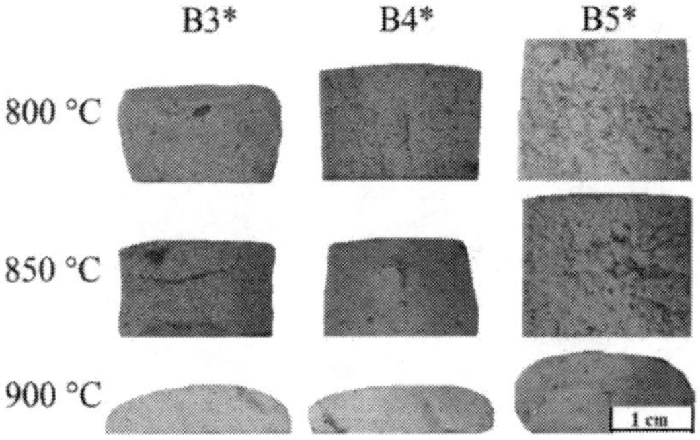

Figure 6. Internal structure of B* series samples, modified with flux.

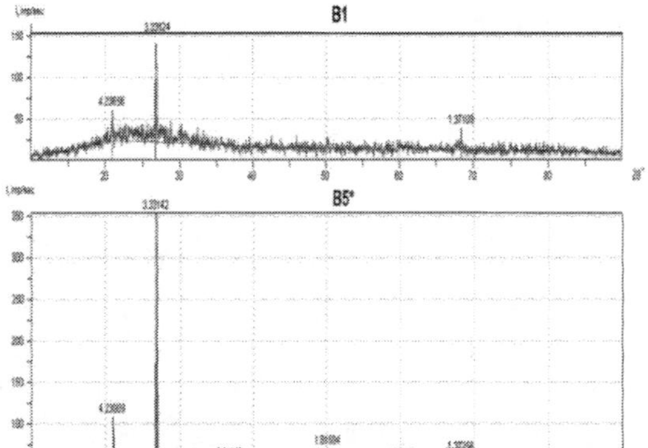

Figure 7. Results of X-ray phase analysis of foam glass

Comparing the samples of the B and the B* series, it can be seen that the introduction of then flux significantly accelerated the melting process. In this case, the presence of more diatomite in the composition led to an increase in the amount of the crystalline phase (β-quartz). Also, the intensity of pore formation decreased, which required the introduction of additives that enhance gas generation.

4 Summary

The possibility of using silicate raw materials of the Far East in the production of glass materials for steel pipeline protection was investigated. It was found that diatomite is more eligible for obtaining enamels than traditional raw materials (quartz sands). Coatings based on diatomite can be obtained at lower temperature, have all required properties and uniform defect-free structure. The use of diatomite in the production of foam glass is complicated by its high melting point, the maximum possible amount of diatomite is 20 wt. %. To introduce a larger amount, additional introduction of intensifying additives is required. With their use, up to 80% diatomite can be used in foam glass.

The use of developed materials will significantly increase the service life of pipelines, reduce the cost of their maintenance, and reduce environmental risks in the surrounding areas. The use of local raw materials will allow the introduction of the proposed technology for integrated pipeline protection at the branch enterprises of the Far East of Russia.

5 Acknowledgements

The work was supported by the Russian Science Foundation, Project № 18-19-00455 «Development of ex protection technology for oil and gas pipelines, operated in the Far East of Russia» in the framework of the 2018 competition «Conducting fundamental scientific research and exploratory scientific research by individual scientific groups».

References

1. J. König, R.R. Petersen, Y. Yue, Ceram. Int., **41**(8) 9793-9800 (2015).
2. H.W. Guo, Y.X. Gong, S.Y. Gao, Mater. Lett., **64**(8) 997-999 (2010).
3. O.V. Kazmina, A.Y. Tokareva, V.I. Vereshchagin, Resource-Efficient Technologies, **2**(1) 23-29 (2016).
4. H. Yin, M. Ma, J. Bai et al., Mater. Lett., **185** 511-513 (2016).
5. P.R. Monich, A.R. Romero, D. Höllen, E. Bernardo, J. Clean. Prod., **188** 871-878 (2018).
6. A.A. Sandulyak, V.A. Ershova, A.V. Sandulyak, E.P. Kurenkov, D.A. Sandulyak, Glass Ceram., 71(5-6) 213-216 (2014.)
7. V. Laur, R. Benzerga, R. Lebullenger et al., Mater. Res. Bull. **96**(2) 100-106 (2017).
8. O.R. Lazutkina, A.K. Kazak, A.A. Temereva, S.O. Nedopolz, Glass Ceram., **63**(3-4) 97-98 (2006).
9. O.R. Lazutkina, A.K. Kazak, A.A. Temereva, S.O. Nedopolz, Glass Ceram., **63**(5-6) 170-171 (2006).
10. S.A. Montayev, N.B. Adilova, S.M. Zharylgapov et al., Int. J. Mech. Eng. Technol., **8**(12) 944-954 (2017).
11. K.S. Ivanov, J. Wuhan Univ. Technol., **33**(2) 273-277 (2018).
12. K.S. Ivanov, S.S. Radaev, O.I. Selezneva, Glass Ceram., **71**(5-6) 157-161 (2014).
13. O.V. Kaz'mina, V.I. Vereshchagin, B.S. Semukhin, A.N. Abiyaka, Glass Ceram., **66**(9-10) 341-344 (2009).
14. Information on: http://www.infomine.ru/files/catalog/62/file_62.pdf
15. L.A. Zemnukhova, A.E. Panasenko, A.P. Artem'yanov, E.A. Tsoy, Bioresources, **10**(2) 3713-3723 (2015).
16. A.V. Ryabova, E.A. Yatsenko, V.V. Khoroshavina, L.V. Klimova, Glass Ceram., **74**(7-8) 282-287 (2017).
17. A.V. Ryabova, T.A. Es'kova, N.S. Karandashova, E.A. Yatsenko, V.A. Smolii, Glass Ceram., **71**(9-10) 327-329 (2015).
18. E.A. Yatsenko, B.M. Goltsman, V.A. Smoliy, A.S. Kosarev, R.V. Bezuglov, Biosci. Biotechnol. Res. Asia, **12** 625-632 (2015).
19. V.A. Smolii, E.A. Yatsenko, A.S. Kosarev, B.M. Gol'tsman, Glass Ceram., **73**(5-6) 219-221 (2016).
20. E.A. Yatsenko, B.M. Gol'tsman, A.S. Kosarev et al., Glass Phys. Chem., **44**(2) 152–155 (2018).

MATEC Web of Conferences **242**, 01017 (2018) https://doi.org/10.1051/matecconf/201824201017
ICAFMC2018

The influence of surface opening cracks on the ultrasonic waves near the surface in concrete members

Wan-Yue ZHENG， Qing-bang* HAN andXUE-ping JIANG

College of Internet of Things Engineering, Changzhou 213022, Jiangsu China

Abstract. Based on the velocities of waves near the surface in concrete, the propagation path and travel time of different waves were calculated, and the influence of crack on these waves was analyzed. Some theoretical conclusions were verified by FEM. The ultrasonic waves were received by array transducers and the velocities of different waves were calculated and distinguished according to wave velocity in the simulation. Simulation results show that when sound wave propagates through the crack, longitudinal waves will arrive at the receiving arrays, and the crack will diffract transverse wave to generate new longitudinal wave and transverse wave by mode conversion, and Rayleigh wave will propagate along the surface (including the crack side wall). Finally, it is found that the relationship between crack depth and amplitude and energy is approximately linearly monotone decreasing by studying the waves near surface with different crack depths.

1 Introduction

With the development of modern society, many engineering like motorway, bridges and dams demand lots of concrete whose bulk is huge. Affected by the leap of temperature, evaporation of water, undesirable construction method, however, it's easy to form surface opening cracks on the surface of concrete, which damage structure. Therefore it's necessary to detect these cracks, and get hold of its depth and width to estimate the damage to the structure and the life of structure. There are many methods used in detecting surface opening cracks in concrete, such as: (1) single ultrasonic flat-measured method; (2) perforating method; (3) Rayleigh wave methods. The perforating method breaks workpiece, and much costs ; the single ultrasonic flat-measured method detect cracks less than 500mm; however, the Rayleigh wave method has these advantages: (1) large energy; (2) not affected by fluid; (3) flexible detection for its propagation character.

Large numbers of researchers all around the world have done research on the detection of surface cracks of concrete. Based on the MFCE(Matched Filtering of Center of Energy), Foo Wei Lee, Kok Sing Lim[1] determine the arriving time of Rayleigh wave. G Hevin[2] have found that the variation of spectral ratios between the transmitted and incident wave were studied as a function of the crack depth. Yangyang[3] has combine time reserval and the Rayleigh wave method, and found that enlarging the cracks depth resulted in an increase in the level of distortion in the reconstructed wave signals, and a higher damage index was obtained. Zhang Zaidong[4] obtained the relationship between slot depth and the absorption frequency on the basis of spectral analysis.

Ultrasonic method has its advantages in detecting surface opening cracks, it's necessary to analyze the influence of the surface opening cracks on the waves near surface including Rayleigh wave. Based on the sound velocity of different wave modes near the concrete surface, this paper will analyz the influence of cracks on the waves near surface, and distinguishes the waves near surface by utilizing receiving array . Simulation results show that longitudinal waves arrived at the receiving array at twice, and the crack diffracted transverse wave to generate new longitudinal wave and transverse wave by mode conversion, and Rayleigh wave propagated along the surface (including the crack side wall), when sound waves near the surface propagated through the crack.

2 Influence on the waves near surface

The waves propagate along the surface are: longitudinal wave, transverse wave and Rayleigh wave. The velocity of longitudinal c_P is fastest , and the velocity of transverse c_R and Rayleigh wave c_S is almost half of that of longitudinal wave. The ratio between c_R and c_S is dependent on the Poisson ratio. Here, the Poisson ratio of concrete is 0.33, and the ratio of concrete is 0.93.

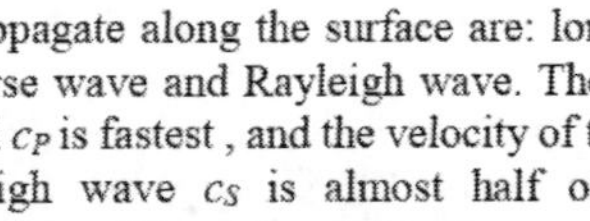

Fig. 1 model geometric structure

On the basis of Fermat theorem, the Snell law near the crack walls and the Huyghens principle near the crack tips, this section analyzes the reflection and diffraction of waves caused by opening cracks, and computes the propagation path and time, and distinguishes the waves finally by velocity. Fig. 1 shows the geometry of the structure, E is excitation source; R is receiving array; A, B are the right and left tips of cracks respectively; C, D are twoe edges of cracks respectively; N, M is the mode conversion point. According theories and geometry, we can obtain:

(1) The longitudinal wave propagated fastest, and the original longitudinal was diffracted near the crack tips, then the diffracted wave propagated to the receiving array at the velocity of longitudinal wave. This phenomenon is used in single ultrasonic flat-measured method[6]. This wave is denoted by 1 longitudinal wave.

(2) According to (1), the original longitudinal was diffracted by crack tips; the velocity of transverse caused by diffraction is slower, so the transverse wave will converse into longitudinal wave by mode conversion when the wave propagated near the surface. This longitudinal wave is denoted by 2 longitudinal wave.

Table. 1 The traveling time and propagation path of different waves.

Wave mode	Propagation time	Propagation path
1 P-wave	$\frac{EA+AB+BR}{c_P}$	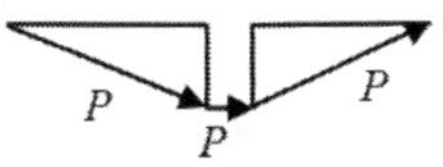
2 P-wave	$\frac{EA}{c_P}+\frac{AB}{c_R}+\frac{BM}{c_S}+\frac{MR}{c_P}$	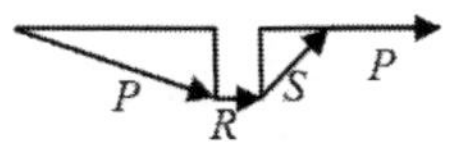
3 P-wave	$\frac{EA}{c_S}+\frac{AB}{c_R}+\frac{BN}{c_S}+\frac{NR}{c_P}$	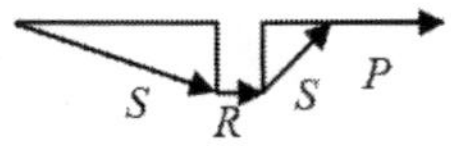
1 S-wave	$\frac{EA}{c_S}+\frac{AB}{c_R}+\frac{BN}{c_S}+\frac{NR}{c_S}$	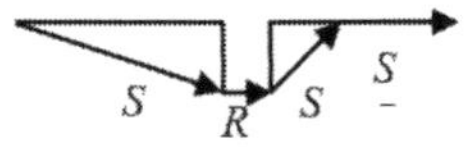
R-wave	$\frac{EC+CA+CB+BD+DR}{c_R}$	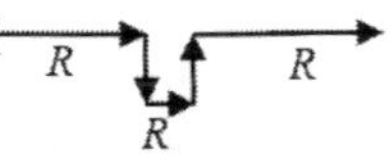

(3) The original transverse wave was diffracted near the crack tips. The transverse wave caused by diffraction propagates into surface, and mode conversion occurs at N point with the introduction of longitudinal wave and transverse wave. Among them, the longitudinal wave propagates at the soonest and is received by receiving array. The corresponding longitudinal is denoted by 3 longitudinal wave.

(4) On the basis of (3), mode conversion at N point introduced transverse wave and Rayleigh wave. Their velocity difference is small for the short propagation path, and received by receiving array. This transverse wave is denoted by 1 longitudinal wave.

(5) As we know, Rayleigh wave propagate along the solid surface., the propagation path of Rayleigh wave was from exciting source to the receiving array including crack walls.

The propagation time and propagation path of different waves mentioned above are presented in Fig1.

3 Numerical model and results

3.1 Numerical model

FEM method was introduced to simulate wave propagation near surface of two-dimension concrete, which is isotropous. Fig.2 shows simulation geometry.

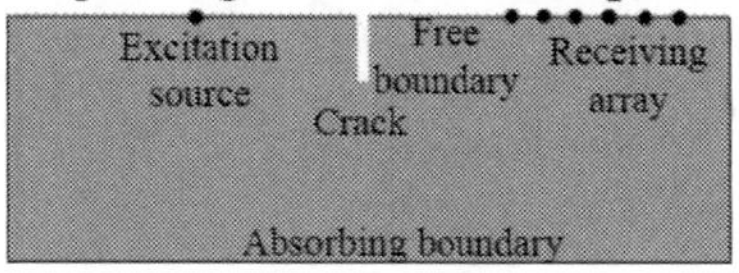

Fig.2 model geometric

The geometry size is 600mm×200mm; filled with concrete; the parameters of material are listed in table 2. Correspondingly, c_P=4013m/s, c_S=2021m/s, c_R=1886m/s.

Table. 2 Medium parameter in the simulation.

Solid acoustic constant	Young modulus	Poisson ratio	Density
value	2.5×10^9[Pa]	0.33	2300[kg/m^3]

In the simulation, biggest size of mesh is 1.67mm; time step is 4.16×10^{-8}s; sampling frequency is 50MHz. Receiving array is adopted in the simulation which are at 400, 420, 440, 460, 480, 500mm denoted as R_1, R_2, R_3, R_4, R_5, R_6 respectively. Contacting surface between concrete and air is almost free boundary, and in order to avoid the disturbance of waves reflected by boundaries, absorbing boundary is adopted. For the sake of near field effect, Rayleigh wave and transverse wave are mixed, thus the distance between transmit-receive probes couldn't be less than 150mm according to document [5]. The distance selected in this paper is 200mm.

There are two simulation models for two simulation purposes, one of which is to study the influence of crack on the waves, and the other is to study the relationship between crack depth and waves near surface, 20, 30, 40, 50, 60, 70, 80mm respectively.

3.2 The influence of crack on the near surface wave

https://doi.org/10.1051/matecconf/201824201017

Based on the parameters listed in table1, sound fields simulated by FEM are seen in figure 3 and 4. By means of observing wave propagation and mode conversion in the sound field, the waves are distinguished received by R1. Figure 5 shows the waves received by R1, and we can find the preceding three waves are longitudinal wave, and the next is slower transverse wave, and the final is Rayleigh wave.

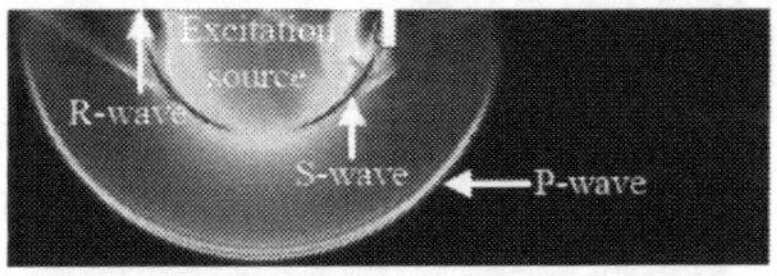

Fig.3 Sound field in 50μs

Fig.4 Sound field in 80μs

To distinguish the received wave modes , the method of one transmitting and multi-receiving mode was adopted in the simulation. Figure 6 shows the wave signals received by R_1 to R_6.

In Fig. 6, line segments were adopted to link the arriving time of different waves for different receivers., the the first three line segments indicate the velocities of longitudinal wave; the last two show transverse wave and Rayleigh wave respectively. Furthermore the relationship between the distance of between receivers and the arriving peak time of different waves was drawn in Fig. 7.

From Figfigure 7, the velocities of every wave were simulated that are: c_P=4013m/s, c_S=2061s, c_R=1886m/s. the velocity of transverse wave is bit faster than the Rayleigh waveIn the sound field, the propagation times of different wave mode were simulated based on the propagation paths and velocities of different wave mode.

Obtained from Figs 8 and 9, the original longitudinal wave was diffracted to introduce the new longitudinal wave and the new transverse wave. According to the Fermat theorem, the traveling time of the new longitudinal wave received by R_1 is:

$$t_{p1} = \frac{(EA + AB + BR)}{c_P} = 52.3\mu s \qquad (1)$$

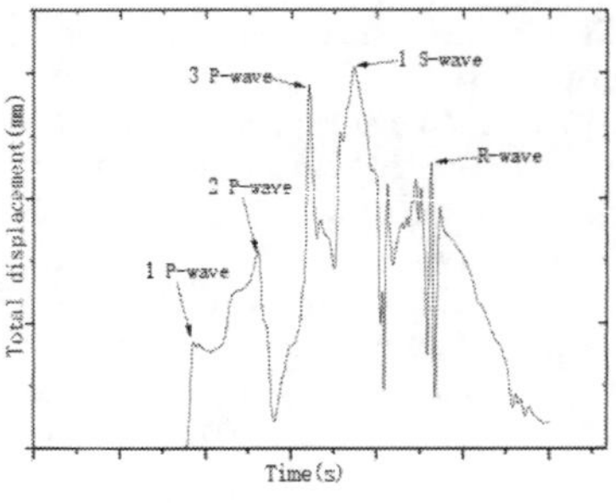

Fig.5 Waves near the surface received by R1 at 400mm

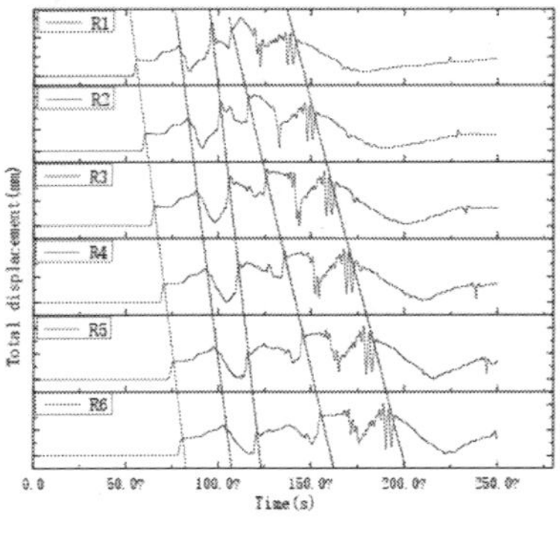

Fig.6 Waves near the surface received by R1, R2, R3, R4, R5, R6

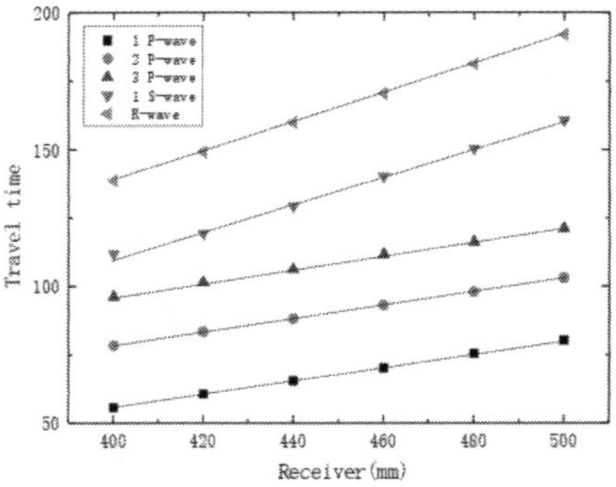

Fig.7 sound velocity of different wave modes in the wave near surface

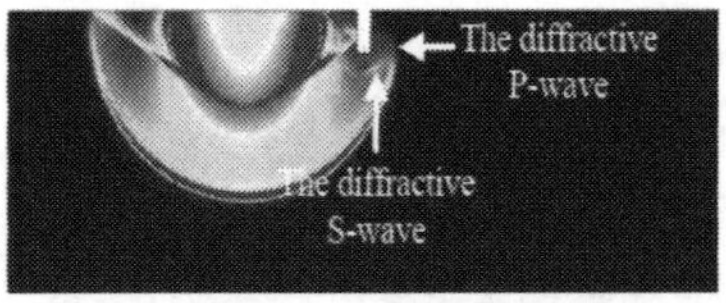

Fig.8 model geometric

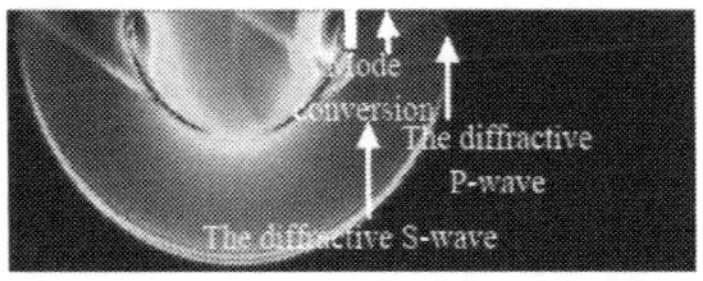

Fig.9 model geometric

The subscript p is on behalf of the longitudinal wave. Obtained from figure 9, the new transverse wave introduced by the original longitudinal wave propagated to the concrete surface at a certain angle, and the mode conversion was introduced. The point of mode conversion is approximately at M (375mm, 200mm) referred to the sound field. Therefore the traveling time of the second longitudinal wave (2 L-wave) received by R_1 is:

$$t_{p2} = \frac{EA}{c_P} + \frac{AB}{c_R} + \frac{BM}{c_S} + \frac{MR}{c_P} = 76\mu s \quad (2)$$

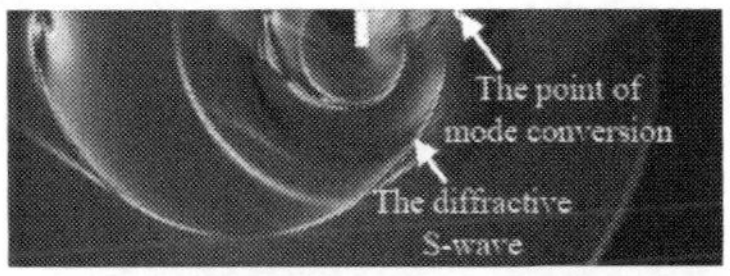

Fig.10 Sound field in 90μs

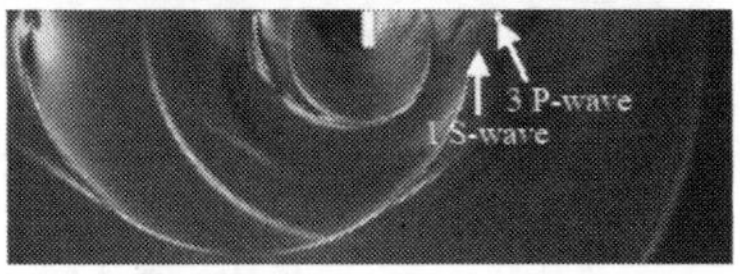

Fig.11 Sound field in 97μs

Obtained from figure 10 and 11, the original transverse wave was diffracted near the cracks tips, and the diffracted transverse wave which is dominating in the waves introduced by diffraction propagated to the concrete surface, with mode conversion similarly. The point of mode conversion is approximately at N (356mm, 200mm) refer to the sound field. Therefore the traveling time of the third longitudinal wave (3 L-wave) and the transverse wave received by R1 is:

$$t_{p3} = \frac{EA}{c_S} + \frac{AB}{c_R} + \frac{BN}{c_S} + \frac{NR}{c_P} = 96\mu s \quad (3)$$

$$t_{s1} = \frac{EA}{c_S} + \frac{AB}{c_R} + \frac{BN}{c_S} + \frac{NR}{c_S} = 106\mu s \quad (4)$$

Finally the slowest Rayleigh wave propagates along the surface including that of cracks only, therefore the traveling time of Rayleigh wave received by R1 is:

$$t_r = \frac{EC + CA + CB + BD + DR}{c_R} = 137.9\mu s \quad (5)$$

The subscript s indicate the transverse wave and the subscript r implys the Rayleigh wave.

The simulation results show that longitudinal waves will arrive at the receiving array at twice, and the crack will diffract transverse wave to generate new longitudinal wave and new transverse wave by mode conversion, and Rayleigh wave will propagate along the surface (including the crack side wall), when sound waves near the surface propagate through the crack.

4 Conclusion

Based on the velocity of waves near surface, Fermat theorem, Snell law near the crack side wall; Huygens principle near the crack tips, ray-tracing algorithm was adopted to analyze the influence of opening cracks in concrete on the waves near surface using finite element simulation. Simulation results show that longitudinal waves will arrive at the receiving array at twice, and the crack will diffract transverse wave to generate new longitudinal wave and new transverse wave by mode conversion, and Rayleigh wave will propagate along the surface (including the crack side wall), when sound waves near the surface propagate through the crack. Simulation results show that with the increase of crack depth, the amplitude of 1 L-wave and the energy of the 1 L-wave and the 2 L-wave decrease approximately monotonously.

References

[1] Lee F W, Lim K S, Chai H K. Determination and extraction of Rayleigh-waves for concrete cracks characterization based on matched filtering of center of energy[J]. Journal of Sound and Vibration, 2016, 363: 303-315.

[2] Hevin G, Abraham O, Pedersen H A, et al. Characterization of surface cracks with Rayleigh waves: a numerical model[J]. NDT & E International, 1998, 31(4): 289-297.

[3] YANG Y, ZHANG M Y, XIAO L, QU W Z. Simulation study of surface breaking crack in concrete beam based on time-reversed theory of surface wave[J]. Concrete, 2015 (12): 17-20.

[4] ZHANG Z D, LU C, WEI Y F. Numeircal simulation to Sizing surface-breaking Cracks by ultrasonic Surface Wave spectral Methed[J]. Nondestructive Testing, 2009 (12): 993-997.

[5] FENG R Y, CHEN Y, LI Z S. Cracks identification of mass concrete structures with R wave spectral energy transmission ratio method[J]. Journal of vibration and shock, 2016, 35(12):221-225.

Application of Python language in UOE molding simulation of pipeline steel

Lian Yeda[1], Zhang Bing[1,a] and Wu Renqiang[1]
[1]*Northwestern Polytechnical University, School of Mechanics, Civil Engineering and Architecture, Xi'an, 710129,China*

Abstract. The ABAQUS plug-in interface based on Python language realizes geometrical design and automatic modeling of gas pipeline UOE molding, which solves the cumbersome problem of manually building complex geometric models. In this study, the algorithm for different sizes of pipelines corresponding to different molds was designed. At the same time, as the ABAQUS kernel scripting program was written, a GUI interface was developed. The interface was used to realize automatic modeling and analysis and control of the calculation work, which laid a solid foundation for practical engineering application analysis.

1 Introduction

1.1 Introduction to Python language

Python is an object-oriented interpreted computer programming language. It was invented by the Dutch Guido van Rossum in 1989. In 1991, he released the first public version of Python. The basic code base provided by the Python language is very rich, covering the advantages of other programming languages at that time. Since its official release, the development history has been more than twenty years, and it has developed to a relatively stable stage. Teng Jun et al. [1] conducted secondary development based on ABAQUS using Python. The interface of Python language exists in finite element analysis software ABAQUS, therefore, through the authoring of the Python language script code and the ABAQUS secondary development interface [2-3], the parameterized automatic modeling process can be realized by combining the GUI interface created by the plug-in toolkit in ABAQUS with the Python language script [4].

1.2 Introduction to UOE molding

The production process of pipeline steel is of great significance for the safe transportation of oil and natural gas. Every country in the world pays great attention to the transportation of oil and natural gas, because the pipelines cross over a long distance and pass through land and sea. Once there is a problem in the transportation process, it will threaten the safety of people and property. UOE is a pipe forming method and is mainly divided into U molding, O molding, and expanding. Figure 1 and Figure 2 show the actual pictures of pipeline steel forming. The efficiency of forming welded pipes is very high and it is widely used in various countries in the world.

[a] Corresponding author: 970427933@qq.com

Figure 1. Pipeline steel pipe in U molding process.

Figure 2. Pipeline steel pipe in O molding process.

2 The development of visual interface

2.1 Introduction to GUI functions in ABAQUS

The finite element software ABAQUS provides some interfaces for secondary development. Therefore, it is possible to use a computing platform to compile a simple modeling program and achieve rapid modeling. The secondary development [5-6] mainly has the following ways:

(1) Some new models can be easily defined by user subroutines to suit their own needs and achieve the purpose of controlling the calculation process and results.

(2) By changing the initial environment, more work can be saved.

(3) The establishment of finite element models and computational analysis of models can be achieved by compiling kernel scripts.

(4) Create a new visual user interface through the compilation of Python language.

2.2 Parameterized design for pipe forming of pipeline steel

It takes about 3 hours to build a simulation using ABAQUS's own graphical interface. However, it has different die sizes for different welded pipe sizes, which greatly increases the modeling time and the workload during simulation. For such a problem, it can be solved by establishing a parametric model. Firstly, the basic parameters of the initial sheet must be determined. As shown in Fig. 3, the relevant parameters of the sheet material are width (W), length (L), thickness (T) and sheet yield strength (q). Based on the parameters of the sheet material, the relevant parameters of the mold can be determined by using the function relationship. Secondly, the positions of the molds and the panels in the molding process are determined. Finally, these basic parameters can be input through a graphical interface for modeling. The establishment of the sheet metal model and the definition of the material are completed, and the corresponding tube molds are also modeled. The entire complex modeling work is completely established by the visualized graphical parameter modeling program. The entire modeling process only takes about 2 minutes.

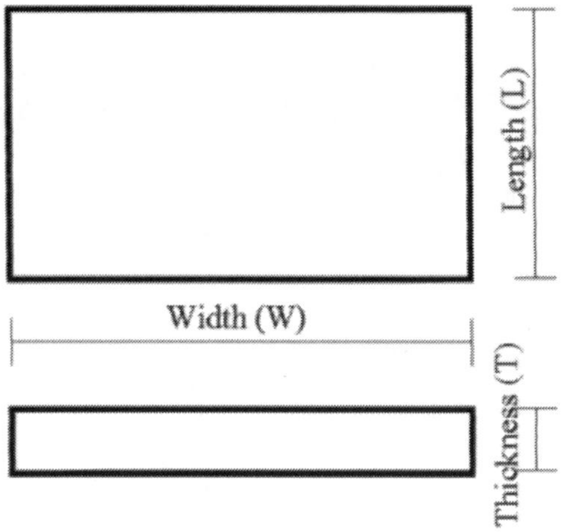

Figure 3. Basic dimensions of sheet.

2.3 The Design of graphical visual interface

Based on the parameters determined in the parametric design of the simulation, this article establishes a visual graphical interface input by simplifying the complex parameters of the model. The process of writing graphical interfaces directly through code functions is pretty complicated. In this process, it not only needs a certain programming basis, but also needs to be very familiar with the functions and application methods of objects. In order to solve this kind of problem, this article chooses a relatively simple and effective method, which is to use the package (Really Simple GUI), which already provides some edited basic controls. Users can create components more conveniently and quickly, and then add the controls that the user needs, and finally create a visual graphical application interface. This article introduces the development process of the visual interface in ABAQUS in more detail. First, after starting ABAQUS/CAE, click on [Plug-ins menu] and select [ABAQUS] from the drop-down menu, then click on the [RSG Dialog Builder] in the menu that continues to appear, and the graphical interface creation dialog box shown in Figure 4 will pop up. In the dialog box, you can create a visual interface and add and modify titles, content, appearance, and functionality. In the process of setting up the development interface, you can click the [Show Dialog] button to pop up the dialog box for the graphic interactive interface that is being edited, which is to display the dialog box that has already been constructed. The establishment of graphical visual interactive interface is mainly accomplished through the following steps:

(1) Change [Title] to [Pipeline Steel Parametric Modeling Program].

(2) Click the [Group Box] button in the left part of the second part of the toolbar and add a group box under the title. The purpose of adding a group box is to facilitate the interface division and make the interface more intuitive.

(3) Click the [Text Filed] button in the fourth part of the left side of the toolbar to add a text edit box. The text label is the content corresponding to the input parameters. The default input content can be set to [Part Name], and you can click [Show Dialog] to view the interface after setting.

(4) Edit the interface in the same way and insert images.

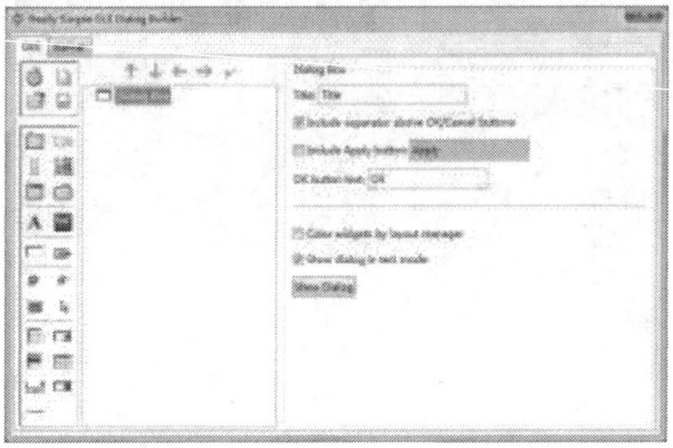

Figure 4. RSG (Really Simple GUI) Dialog Builder.

After the establishment of the user graphic interactive interface is completed, the corresponding Python script file needs to be selected as a module name on the Kernel tab page. Select the script file written in Python, and then select the corresponding function expression relationship, and connect the dialog and ABAQUS kernel script through the "module. Name. function. Name" command. After the interface is developed, you can directly save the plug-in program. After the interface is developed, you can directly save the plug-in program. All relevant file information contained in the plug-in program is stored in the ABAQUS plug-in directory.

2.4 The development of kernel script

This article uses a relatively quick way to develop a kernel script. When performing various operations in ABAQUS/CAE, the background program will be recorded in the "Abaqus.rpy" file in the current working directory in real time. In addition, when each time the CAE model is saved, a .jnl file with the same name as the model name is automatically saved in the directory where the file is saved. The script commands used in the entire modeling process are recorded in files whose suffixes are ".rpy" and ".jnl". The above two types of files are opened by using text editing software, the contents of the script records are analyzed, and the most optimized instructions are extracted according to their own requirements. Then modify the instructions, edit the corresponding function with the visual interface, you can quickly edit the kernel script. In this way, we can save a lot of script editing time and improve the efficiency of scripting by users.

 https://doi.org/10.1051/matecconf/201824201018

3 A case study of interactive interface

First open the plug-in program in ABAQUS, as shown in Figure 5. The size of the sheet material chosen in this paper is width W=1219 mm, thickness T=33 mm, and length L=600 mm. The material of the sheet is X80 pipeline steel with a yield strength of 445 Mpa. First input the corresponding parameters into the dialog box, and then click the OK button to begin the establishment of the finite element model. Figure 6 is a screenshot in the automatic modeling process. The automatic modeling essentially establishes operation commands through plug-ins, and then quickly builds the model.

It takes only about 2 minutes to build the model and build all the components. Figure 7 and Figure 8 show the assembled model. The workload of building models in the interactive interface is greatly simplified compared to that in ABAQUS, and at the same time it provides convenience for researchers in practical projects.

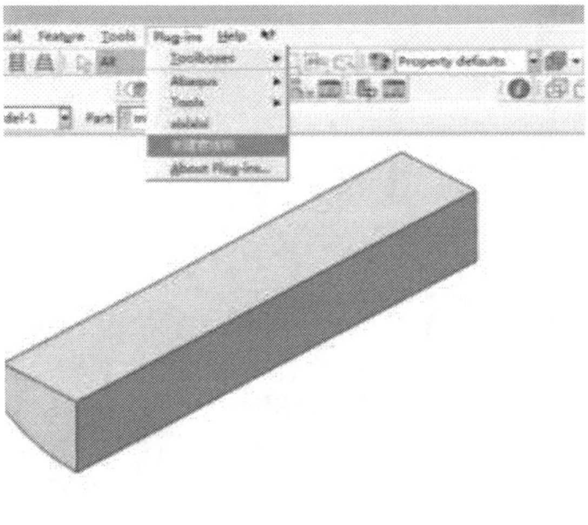

Figure 5. Open the ABAQUS plug-in.

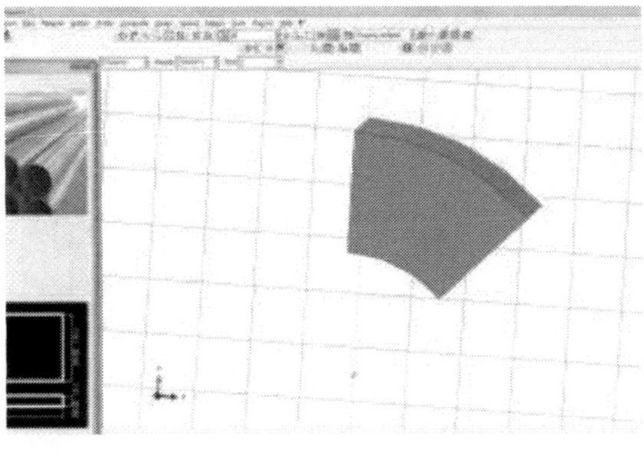

Figure 6. Automated modeling process.

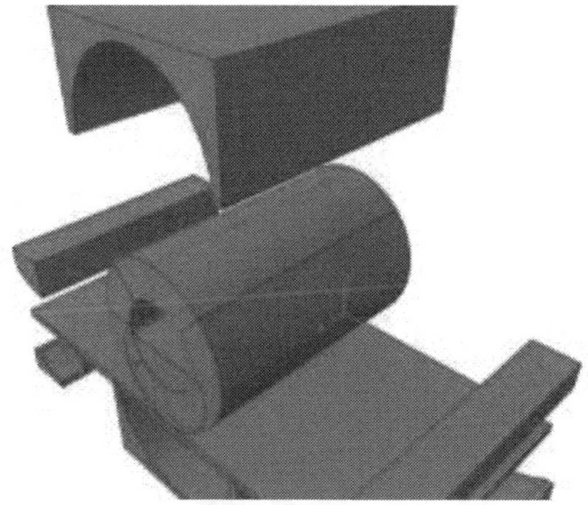

Figure 7. 3D view after modeling.

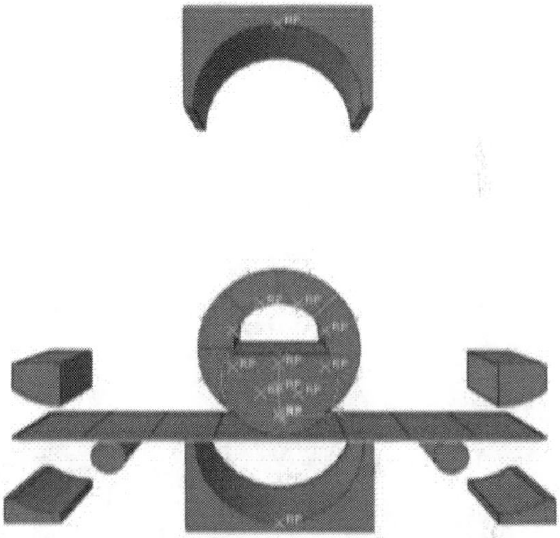

Figure 8. Front view after modeling is complete.

4 Conclusion

In this paper, a visual graphic interface is established by writing a Python language script, and parametric modeling of the sheet and mold in the UOE molding process is realized. For the problems of complex model and difficult modeling, the modeling method of visual graphical interface can be used to solve it, and the model can be established quickly and automatically. Whether in scientific research or in actual engineering, the developed visual interface can effectively reduce working hours, improve work efficiency, and provide assistance for engineers and scholars.

References

[1] J. Teng, H. Zhang, Z.H. Li. Journal of Disaster Prevention and Mitigation Engineering, **33**,9 (2013).
[2] R.M. Xie, X.J. Wang. China Water Transport, **17**, 77 (2017).
[3] X.D. Zhang, J. Liu, S.G. Zou, W.Y. Tang. Journal of Plasticity Engineering, **24**, 140 (2017).
[4] H. Wang, T.C. Li, Q.B. Wang, Y.B. Zhan. Modern Tunnelling Technology, **52**, 160 (2015).
[5] Q. Zhang, Y. Ma, S.C. Li. Ship Electronic Engineering, **31**, 131 (2011).
[6] H.W. Li, W.D. Wang. Journal of Building Structures, **34**, 353 (2013).

Design and trial-produce of a special nail-tooth thresher for pulses crops

Xianfei Xia, Qiaomin Chen*, Hongru Xiao and Guang Yang

Nanjing Research Institute for Agricultural Mechanization, Ministry of Agriculture, Nanjing 210014, China.

Abstract. Pulses crops are widely distributed in China, but mechanization production-level of this crops is relatively low at present. So, a special nail-tooth thresher for pulses crops was carried out in this project to solve the problem of mechanical threshing. The roller diameter, speed, length, and nail-tooth structure, stripper concave screen, power device and cleaning system are determined by detailed theoretical calculation. Testing result of the trial-production prototype showed that the threshing roller with large diameter and low rotating speed, the nail-tooth threshing mechanism and the concave plate screen with large spacing are suitable for the broad bean thresher. The optimal combination is roller speed 400 /min, feed quantity 1.2 kg/s and concave tolerance 54 mm. Testing results showed that impurity content and breaking rate of the developed nail-tooth type broad bean thresher were respectively 1.09 % and 3.45%, It could improve the efficiency of the pulses crops threshing process significantly.

1 Introduction

Pulses crops are widely distributed in China, there are more than 20 kinds that have been cultivated, and the main varieties are mung bean, adzuki bean, broad bean, pea and kidney bean. The pulses crops have the function of improving the conditions of soil nutrient and feeding for draught animals. It is an ideal crop for grain-legume rotation and organic dry farming, and has important significance for the sustainable development of agriculture. Broad bean is a representative variety of pulses crops.

Broad bean, also known as Buddha bean, LuoHan bean, orchid bean and so on. With a protein content of about 30%, broad bean is nutritious and widely used. It is one of the important legume crops in the world, and broad bean is mainly distributed in northwest, southwest and Yangtze River Delta areas in China. The planting area is about 800,000 hectares (12 million mu), and the total annual production is over 1.6 million tons.

However, mechanization production-level of the broad bean is relatively low at present, and most of them are completed by artificial harvesting (or pod picking), threshing and cleaning. The operation intensity is large and the production cost is high. For the traditional corn or rice thresher, its threshing operations primarily through the installation of the tooth with high-speed rotation on the spindle, is not suitable for pulses crops. So, the demand for mechanized harvesting equipment is very urgent. Due to the large grain size and high straw specific gravity, broad bean is easily broken during mechanical threshing, and the impurities are hard to clean. Therefore, performance requirements of the machine are relatively high [1-3]. In this context, the design and development of broad bean thresher was carried out in this project to solve the problem of mechanical threshing in the broad bean harvesting process.

2 Design of the main components

Design parameters of the main components include roller diameter, roller speed, nail-tooth structure, length of the roller, stripper concave screen, power device and cleaning system. They are determined by the following methods.

2.1 Roller diameter

Diameter of the roller has an important influence on threshing effect, when the diameter is too small, straw is easily entangled and the separation area of the concave plate is also directly reduced. Large diameter threshing roller can effectively increase the threshing and separating time of ears, and improve the threshing and separating ability and production efficiency. At present, the commonly used diameter of rice and wheat threshing machines is 450~650 mm, then roller diameter of the threshing machine studied in this paper is selected as 400mm through comprehensive analysis [4].

2.2 Roller speed

The nail-tooth threshing mechanism realizes the separation of grains and stalks by the action of the hammer [5-9], line speed range at the end of the nail is usually 7~13 m/s. However, due to the large grain size of broad bean, the linear speed of roller cannot be too high, otherwise it is easy to damage broad bean in the threshing process. While the operation efficiency will be affected if

Corresponding author: dire-chen@foxmail.com

it is too low. Therefore, 9 m/s is selected. So, according to formula 1, the speed of roller is 429.93 r/min.

$$n = \frac{60v}{\pi D} \quad (1)$$

Where, D is the diameter (mm) of the roller; v is the linear speed (m/s) of the nail-tooth.

2.3 Nail-tooth structure

The number and arrangement of the nail-tooth has a direct influence on the threshing performance indicators of un-threshing, entrainment loss and broken grain rate. It determines the work quality and productivity of the machine [10].

Design of the nail teeth is arranged evenly in 4 rows and fixed on the roller, the nail tooth height is 66 mm, diameter is 12 mm, tooth distance is 100 mm and the number of teeth is 40 in this project. In addition, the rubber layer is covered on the teeth to reduce grain broken. When the broad bean straw enters the threshing chamber through the hopper, the nail teeth rotate and agitate the broad bean straw. On the one hand, the roller has a tangential strike force along the cylinder circumference, on the other hand, it has a right thrust along the axial direction, forcing the material to move toward the discharge port.

2.4 Length of the roller

The length of roller threshing section is an important parameter affecting threshing quality, if the roller is too short, the net removal rate will be affected. While the roller is too long, it will increase the manufacturing cost and the size of the threshing machine. The length of roller is calculated as follows.

$$L = a(\frac{Z}{K} - 1) + 2\Delta L \quad (2)$$

Where, a is the distance between teeth; Z is the total number of teeth; K is the number of teeth rows; ΔL is the distance from the nail teeth to the end of the tooth rod, generally 15-25 mm.

The length of the roller is 950 mm calculated by equation 2.

2.5 Stripper concave screen

In the threshing process, the concave plate can well isolate the straw, and at the same time, the screen holes distributed on the concave plate can make the shedding broad bean seeds leak to the clear screen [11,12].

There are two main types of concave structure: grid and punch, and the grid concave has great strength, good rigidity, large sieve hole and good threshing and separating effect. The threshing process of broad bean is mainly completed in the front part of the roller, and the latter part is mainly used for separating and threshing, so the grid form threshing concave plate is adopted.

The screening efficiency of concave screen depends mainly on the length and effective separation area of the concave plate. In a certain range, screening efficiency will increase with the length of the concave screen. At the same time, the larger hole ratio of concave plate screen, the higher screen efficiency of concave screen. The hole ratio of concave plate screen A is calculated by equation 3. The hole ratio of concave plate scree of the machine developed in this paper is 72 %.

$$A = \frac{a_l b}{r\pi L} \times 100\% \quad (3)$$

Where, A is the hole ratio of concave plate scree, a_1 is the number of sieve holes, b is the area of sieve hole, r is the roller radius, L is the roller length.

2.6 Power device

The thresher shall be equipped with motor to provide threshing power, screening power and cleaning fan power. Combined with the structure design and practical application of the thresher, the matching motor with 5.5 kW power is selected under the condition that the threshing capacity is 3 t/h and the thresher works normally.

2.7 Cleaning system

From feeding broad bean material into the threshing section, to obtaining clean bean seeds, a large amount of straw, dust and sundries mixed with materials should be cleaned by the cleaning system. The separation principle is as follows: the fallen bean seeds and small impurities fall through the grid concave hole, during the falling process, the fine impurities are sucked into the fan by the negative pressure fan and discharged from the fan outlet while the grains fall into collection box with the action of reciprocating vibrating screen.

3 Entire design and test verification

The entire design was developed by integrating the above design parameters as shown in Figure 1, then the prototype production was carried out. In order to verify the working performance, the full-scale experiment was finished and the key operating parameters were tested. The operating principle and testing methods of impurity rate and breaking rate are as follows.

3.1 Operating principle

The naturally dried straw after cutting was sent into threshing machine from feeding mouth, motor drive the roller to rotate at a certain speed, then the fast-moving nail keep hitting the straw in the threshing chamber to achieve the rapid separation of grains and straws. Next, grain entered the cleaning system through sieve mesh gap, and the remained straw was discharged out of the machine through crushing device. The clean grain can be obtained at the discharge port.

The main indexes to evaluate threshing machine operation performance include breaking rate and impurity rate.

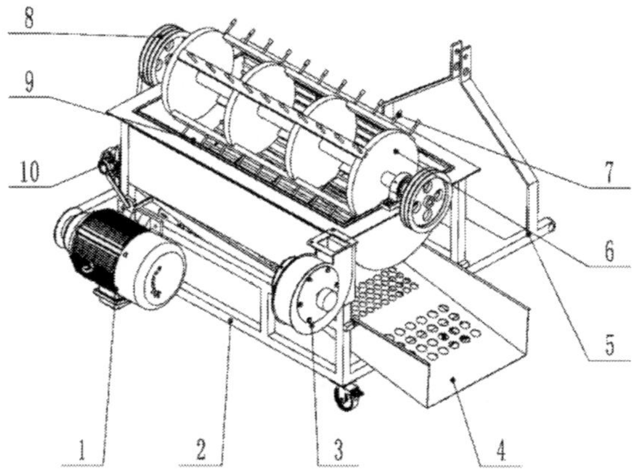

1-motor, 2-frame, 3-cleaning fan, 4-grading screen, 5-suspension frame, 6-threshing roller, 7-nail-tooth, 8-driving wheel, 9-gravure screen, 10-vibration mechanism

Figure 1. The overall structure diagram of broad bean threshing machine.

3.2 Breaking rate

According to the standard requirements, a random sample of broad bean no less than 2000 g was collected and the seeds with injuries and obvious cracks were selected out and weighted. The breaking rate is calculated according to equation (4).

$$Z_s = \frac{W_s}{W_i} \times 100\% \tag{4}$$

Where, Z_s is grain breaking rate, %. W_s is the mass of broken grains, g; W_i is the total grain mass of the sample, g.

3.3 Impurity rate

After each testing, 2,000g of broad beam seeds were collected from the outlet. Impurities and grains were separated and weighed separately. The impurity rate is calculated according to equation (5).

$$S_W = \frac{W_w}{W_z} \times 100\% \tag{5}$$

Where, S_W is grain impurity rate, %. W_w is the mass of impurity, g; W_z is the total mass of the sample, g.

3.4 Experimental design

The controllable conditions in the experiment are feed quantity, concave tolerance and roller speed. The range of each testing variables is shown in Table 1. Different experiments are arranged by the orthogonal experiment design method. The experiment is adopted $L_9(3^4)$ orthogonal tables to repeat 3 times. The correlation of variables is analyzed and the optimized parameter combinations were obtained.

Table 1. Experimental factors and levels.

Levels	Roller speed r/min	Feed quantity kg/s	Concave tolerance mm
1	400	0.75	50
2	450	1.00	54
3	500	1.25	58

3.5 Testing result

According to the above methods, threshing performance of the prototype was tested. Variety of the testing broad bean was Qidou 2, and moisture content of the bean straw was 12%-16%. Testing results was shown in Table 2, Range analysis of the experimental results was shown in Table 3.

Through range and variance analysis, we found that each factor has different effects on the variables, the roller speed has the greatest impact on the breaking rate and. The primary and secondary sequence of the factors affecting breaking rate is roller speed > feed quantity > concave tolerance, while the optimal combination is roller speed 400 /min, feed quantity 1 kg/s and concave tolerance 54 mm. In addition, the concave tolerance has the greatest impact on the impurity rate and. The primary and secondary sequence of the factors affecting breaking rate is concave tolerance > roller speed > feed quantity, while the optimal combination is roller speed 400 /min, feed quantity 1.25 kg/s and concave tolerance 54 mm.

It can be found that the optimal combinations are significantly different when a certain optimization variable is chosen. After weighing the influence of impurity rate and breaking rate on the result, the final optimal combination is roller speed 400 /min, feed quantity 1.2 kg/s and concave tolerance 54 mm. Testing results under this combination showed that impurity content and breaking rate were 1.09 % and 3.45% respectively, which met the requirements of application. The threshing effect of the machine is shown in Figure 2.

Table 2. The orthogonal experiment arrangement and results

Experiment No.	Experiment factors			Result	
	Roller speed (r/min)	Feed quantity (kg/s)	Concave tolerance (mm)	Breaking rate (%)	Impurity rate (%)
1	1	1	1	3.25	0.5
2	1	2	2	2.79	0.36
3	1	3	3	3.27	0.2
4	2	1	2	3.86	0.32
5	2	2	3	3.13	0.47
6	2	3	1	3.81	0.65
7	3	1	3	3.91	0.68
8	3	2	1	3.65	0.45
9	3	3	2	3.42	0.31

Table 3. Range analysis of the experimental results.

Evaluating index	Range	Roller speed (r/min)	Feed quantity (kg/s)	Concave tolerance (mm)
Breaking rate	k_1	3.103	3.673	3.57
	k_2	3.6	3.19	3.357
	k_3	3.66	3.5	3.437
	R	0.557	0.483	0.213
Impurity rate	k_1	0.353	0.5	0.533
	k_2	0.48	0.427	0.33
	k_3	0.48	0.381	0.45
	R	0.127	0.113	0.203

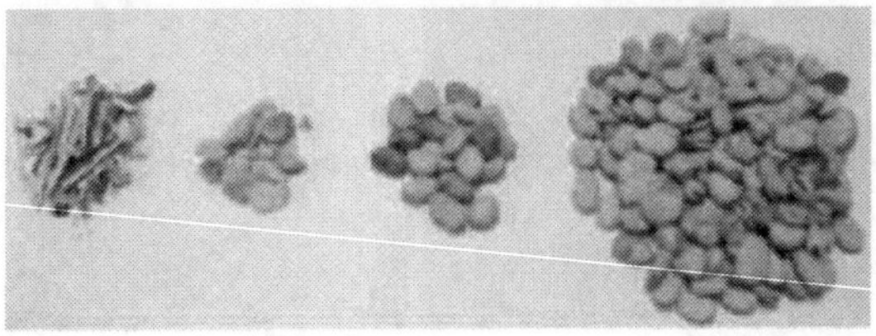

(a) impurities (b) the damaged grain (c) the ideal grain
Figure 2. Threshing performance of the prototype

4 Conclusion

(1) In order to improve efficiency of the broad bean threshing process, the threshing roller with large diameter and low rotating speed, the nail-tooth threshing mechanism and the concave plate screen with large spacing are suitable for the broad bean thresher.

(2) The optimal combination is roller speed 400 /min, feed quantity 1.2 kg/s and concave tolerance 54 mm. Testing results showed that impurity content and breaking rate of the developed nail-tooth type broad bean thresher were respectively 1.09 % and 3.45%, achieved relatively ideal operation effect.

Acknowledgments

This work was financially supported by China Agriculture Research System (CARS-08).

References

1. Hongz T, Bin T, Xiaoh T and Ming L, Chemicophysical properties of twenty broad bean varieties and their correlations, J. the Chinese cereals and oils association. 24 (2009) 153-157.
2. Xiaoh T, Hongz T, Bin T, Ming L, and Yangx L, Analyse physicochemical properties of broad beans in main producing areas in China, grain engineering. 17(2009) 7-11.
3. Mianc J, Dongx X, Yunq G and Hongx R, Simplified cultivation techniques for high yield and high efficiency of faba bean in northern hebei province, agricultural science and technology newsletter. 8 (2014) 216-217.
4. Gang W, Zhuoh G, Senl M, Qing T, Chongy W,optimization of operating parameter and structure for seed thresher device for rape combine harvester, Transactions of the CSAE. 33(2017) 52 – 57.
5. Fengj W, Hongj Z, and Shig W, research and development of the single-drum nail-tooth corn thresher, Journal of Qiqihar University. 31(2015) 26-29.
6. Lianx G, Shiy Z, Ruix C and Dex Y, design and experiment on soybean breeding thresher of double feeding roller and combined threshing

cylinder, Transactions of the CSAM. 46(2015) 112-118.

7. Lianx G, Shiy Z, Ruix C and Dex Y et al, development and performance test on pneumatic cleaning-circulatory device of soybean thresher, Transactions of the CSAE.28 (2012) 22-27.
8. Lizhang X, Yaoming L, Chenghong W, et al, a combinational threshing and separating unit of combine harvester with a transverse tangential cylinder and an axial rotor, Transactions of the CSAM. 45(2014) 105-108.
9. Yaoming L, Wei Z, Lizhang X, et al, parameter test and optimization of tangential-horizontal-horizontal threshing and separating device, Transactions of the CSAM. 46(2015) 62-67.
10. Yaoming L, Tao S, Lizhang X, performance test and analysis of rape multi cylinder threshing and separating device,Transactions of the CSAE. 29 (2013) 36-43.
11. Zhifeng D, Zhongkai C, Hua Z and Jin Z, design and experiment of rasp bar and nail tooth combined axial flow corn threshing cylinder, Transactions of the CSAE 34(2018)28 -34.
12. Zhongkai C , Zhifeng D, Zhoujin, et al, design and research on 5TYS280 corn threshing and cleaning test bench, Journal of Agricultural Mechanization Research. 39(2017)113–117.

MATEC Web of Conferences **242**, 01020 (2018) https://doi.org/10.1051/matecconf/201824201020
ICAFMC2018

Timber "Science" in Herculaneum during the 1st Century A.D.

Nicola Ruggieri[1,a]

[1]*Soprintendenza Archeologia Belle Arti e Paesaggio per le province di Catanzaro Cosenza e Crotone, piazza Valdesi, 13 - Cosenza - Italy*

Abstract. The paper deals with the state of the knowledge on wood material used for constructions during the 1st century in Herculaneum, focusing in particular on the analysis of durability measures and for limiting the effects of timber shrinkage. The study founds on the historical literature and on the finds exceptionally conserved in the archaeological site "thanks" to the Vesuvius' eruption.

1 Introduction

The theme of the conservation of ancient constructions belonging to the historical and architectural heritage and in particular ruins in archaeological sites [1-4] is more and more generating interest in researchers to develop investigation methodologies [5-11] and innovative technologies for strengthening them [12-15]. In this field the paper presents the results, relate to the material characterization, regarding the analysis carried out on timbers discovered in the archaeological site of Herculaneum. The terrible Vesuvius' eruption of the 79 A.D. that buried Pompeii, Herculaneum and Stabia was characterized by a Plinian eruptive dynamics, with different effects on the hit towns relate to the orientation, the buildings disposition and the distance from the volcano. Specific studies on the deposits stratigraphy discovered in Herculaneum reveals a sequence that was dominated by pyroclastic flows alternated to surges. The high temperatures reached by the volcano's winds, till approximately to 400 °C, provoked the wood combustion. Immediately after the deposit of eruptive materials, the consequent generation of an environment lacking of oxygen [16] entailed the transformation of the wood in a chemically stable material devoid of organic components, therefore not degradable by biotic agents. Others archaeological finds, as those of the *Casa del Rilievo di Telefo* (fig. 1), seem to have been inundated by mud flows (lahar), constituted by pyroclastic debris and water [17] which, covering the wooden material still not charred, generated a preserving environment from biotic decay, allowing its conservation until the discovery at the end of the 20th century.

It is an exceptional condition that, with the opportune aid of the historical sources, has permitted to study the material features and the installation modalities of the carpentries in Herculaneum during the 1st century, from which to deduce the knowledge advancement on timber material.

[a]Corresponding author: nicola.ruggieri@beniculturali.it

Figure 1. Herculaneum, *Casa del Rilievo di Telefo.*

2 Material characterization

2.1 Defects

The strength parameters have an extreme variability in dependence of the material features relatively, in particular, to the grain inclination and the position and dimension of the knots, besides the timber specie. The palynological analyses carried out in Campania (Italy) reveal the presence of a forestation surface on Mount Vesuvius with more trees species than in the current situation [18]. The forest of the Lattari's mounts and the Vesuvius were predominantly characterized by arboreal species classifiable as *Quercus pubescens*, *Pinus pinea* and, at altitudes above 200 meters quotes, birch (*Betula sp.*) and silver fir (*Abies alba*) [19]. Furthermore the presence of the hornbeam (*Carpinus betulus*), chestnut *(Castanea sativa),* and trees belonging to the *Populus*

https://doi.org/10.1051/matecconf/201824201020

Figure 2. Herculaneum, *Casa del Rilievo di Telefo*'s roof. The find n. 9 (photo property of HerculaneumConservation Project: Parco Archeologico di Ercolano, Packard Humanities Institute, Istituto Packard per i Beni Culturali).

genus [20] is attested. Therefore, it is a considerable forestry heritage, which arboreal species is coherent with the wooden finds discovered in the Vesuvius' area, that testify an use of poplars, chestnut and oak as well as of silver fir in the construction for the load bearing structures. The latter is very common and represents the wood specie used for the *Casa del Rilievo di Telefo*'s roof [21] discovered during the excavation conducted in the Eighties and expanded between 2009 and 2010. The dendrochronological analysis, based on the master curves made from buildings of the Central Germany that finds correspondence also for the South Italy, dates the wooden elements back to the early years of the 1st century A.D. [22]. Such members, approximately 500, vary both with regard to the conservation state and the original function as well as the dimension and the geometry; some ones, that are no longer intact, are subdivided into two or more elements. The majority of the timbers are characterized by an approximately circular section still with the bark or just un-barked, for which the use as purlins can be hypothesized. The members, considering the conservation situation, water mixed to pyroclastic materials, although most of them are not charred, present an extended and intense cellular degradation with a high percentage of water (MWC). That condition does not permit to detect extended data on the material features and, in particular, on the presence of possible defects that influence the timbers structural behaviour. However, from a visual analysis, the presence of checks due to the shrinkage can be observed, that are useful to provide information on the grain orientation that seems to be regular and almost straight (i.e. finds 9 and 11) (fig. 2).

Figure 3. Herculaneum, The *Collegio degli Augustali*. The "recreation" of the beam, applying erratic carbonized wood parts on a metallic core.

The knots are, in general, not recurrent, with a limited diameter and placed approximately in the middle of the member (for example find 11). Only in few cases the knots are at the lower edge of the beam where, in general, fibers are stretched and (find 8) could entail a fragile rupture. The squared members with a prevalent rectangular geometry, some with bevels, are characterized, most likely, by a saw manufacturing. The thickest elements contain the heartwood almost entirely, a condition that inhibits the timber twisting due to the shrinkage and, concurrently, reduces the cut of the fibres. A more difficult interpretation is that referred to the carbonized wood of Herculaneum, in many cases still in the original position. This analysis gets complicated due to some inadequate intervention of restoration carried out in the 20th c, even if this circumstance is limited only to few finds. In fact, some *anastylosis* interventions have included the "recreation" of the beam, applying erratic carbonized wood parts on a metallic core, generating an alteration of the original position of the wood, hence an anomalous morphology of the material (fig. 3).

2.2 Measures for a proper seasoning and shrinkage control

Timber tends, immediately after the tree knocking down, to lose the water present in the liquid state and as component of cell walls aimed at reaching the equilibrium with the hygrometric conditions of the environment in which it is used. A material moisture variation that causes dimensional modification and different deformations according to the radial, tangential and longitudinal direction due to a strongly anisotropic behaviour of the wood. Several Latin authors [23] provide indications on how to conduct a correct seasoning for controlling and reducing such deformations. Pliny (Naturalis Historia, 77 A.D.) suggests to put the timber under the wheat in order to trigger a slow process of equilibrium between the wood moisture and the external humidity (NH, XIII, 99). Cato (De Agri Cultura, c.a. 160 B.C.) recommends, as a proper methodology, to immerse the beams in a mix of water and manure (cat. Agr. Cult. 31). The method proposed by Columella (De Re Rustica, ca. 60–65 A.D.) is particularly advanced, precursor of the modern exsiccation system; in fact, it suggests to adopt a *fumarium* for the timber seasoning, in which to flow in the heat deriving from the heating of the villa's baths [23]. Therefore, it is an expedient for improving the strength and durability but mostly aimed at inhibiting a scarce dimensional stability that could entail negative interactions with the other materials and, in general, with the construction itself. An use of the continuous boards that cross the entire masonry section, with the aim of obstructing dimensional variations due to the shrinkage, is attested in several floors of Herculaneum. An arrangement that clearly echoes Vitruvius'entry in the *liber* VII, I, " ... no wall has to raise up to the top of the house ... but rather has the board ... In fact if a wall is erected without interruption, when the beams start to dry out and to twist, it, remaining unmovable due to its robust structure, inevitably provokes cracks in the paving on the right and left part ..." ("... ne qui paries, qui non exeat ad summum … habeat coaxationem … Cum enim solidus exit, contignationibus are scentibus aut pandationes identibus, permanens structurae soliditated extra ac sinistra secundum se facit in pavimentis necessario rimas …"). Similar precept was applied in the *Collegio degli Augustali*'s atrium, in which the discontinuity of the masonry has represented a preferential rupture way triggered by the horizontal action determined by the eruption winds, with the consequent generation of a horizontal hinge exactly in correspondence to the wall interruption (fig. 4a).

Figure 4. a) Herculaneum. The *Collegio degli Augustali* atrium; **b)** Herculaneum. The shop belonging to the *Casa del Colonnato Tuscanico*.

2.3 Measures for the durability

The inter-storey floors are, analogously to the Pompeii [24] ones, characterized by unique order of beams arranged with a small spacing, approximately 45 centimetres, with an evident will of obtaining resistant and in horizontal plane stiffen floors. Bricks are present close to the support and they frame the beam socket (fig. 4b). The probable purpose of such an arrangement is to isolate the carpentry from the masonry, possible water carrier, namely to avoid favourable condition for the decay due to biotic factors. However, those fictile elements, most likely also with the function of squaring, in a fast and economical manner, the hole in the chaotic masonry, represent a measure of doubtful effectiveness. In fact the highly porous bricks, rather than to provide isolation, can transfer humidity from the masonry to the timber members, due to capillarity phenomena. Conversely, the roughness generated by a hole execution with not planar surfaces would have achieved limited contact points between the wall and the timber and a consequent beam socket airing.

The carbonized beam of the porch in front of the *Bottega ad cucumas* is peculiar. It is a unique continuous member on three supports, with a static scheme that reduces the bending moment and adequately responds to failures deriving from unequal foundation settlements. That beam is constituted by two overlapping and outdistanced elements. An organization that finds possible references in Vitruvius and, in particular, in his provision regarding the *trabes compactiles* of the Etruscan temples (Liber IV, 7, 4) "… the gap (between the two elements) has to be two inches because if the beams touch each other and are not ventilated, they immediately heat up and quickly rot…" ("… ut compactura duorum digitorum habeant laxationem. Cum eniminter se tangunt et non spiramentum et perflatum venti recipiunt, concale faciuntur et celeriter putrescunt …"). A passage that emphasizes an extraordinary intuition of the Architect of Augustus, even though he makes the mistake of attributing to the heating the cause of the rottenness. In fact, the presence of a ventilation layer between the two overlapping members permits to prevent the condense generation that could imply possible fungi and insects infestation.

3 Conclusions

The devices adopted in the timber carpentries of the ancient Herculaneum testify a deep knowledge of the *materia*. An empirical approach and the precepts contained in the Latin treatises, on which *De Architectura* stands out, constitute the basis of the progress of the timber science during the 1st century in Herculaneum. A particularly advanced knowledge that allowed choosing constructive solutions capable of ensuring the durability and mitigating the effect of the dilatation and contractions caused by the moisture variations and, contemporaneously, capable of permitting to use very strong wood. However, the latter property is, most likely, to be ascribed only to the carpentry of wealthy families dwellings and public buildings.

References

1. N. Ruggieri, S. Galassi, G. Tempesta, Pompeii's Stabian Baths. Mechanical behaviour assessment of selected masonry structures during the 1st century seismic events, Int. J. Archit. Herit. **12**, 5, 859-878 (2018) 10.1080/15583058.2017.1422571
2. S. Galassi, L. Dipasquale, N. Ruggieri, G. Tempesta, Andalusian timber roof structure in Chefchaouen, Northern Morocco: construction technique and structural behavior, J. Archit. Eng. **24**, 3, 04018012 (2018) 10.1061/(ASCE)AE.1943-5568.0000315
3. S. Galassi, N. Ruggieri, L. Dipasquale, G. Tempesta, Assessment of the Moroccan Vernacular Timber Roof: a Proposal for an Eco-Friendly Strengthening System, J. Architectural Conservation (2018)10.1080/13556207.2018.1545105
4. M. Paradiso, S. Galassi, S. Benedetti, Il caratteristico Forte di San Fernando de Bocachica, Colombia, in: Luis Villegas, Ignacio lombillo, Clara Liano, HaydeeBlanco (Eds.), *Patologia della Costruccion, Tecnologia de la Rehabilitacion y Gestion del Patrimonio*, Proc. Conf. Latinoamericano Patologie del costruito, tecniche di riabilitazione e gestione del patrimonio (Rehabend 2014), Santander, Spain, GráficasIguña, S.A. (2014)
5. S. Galassi, N. Ruggieri, G. Tempesta, A novel numerical tool for seismic vulnerability analysis of ruins in archaeological sites, Int. J. Archit. Herit. (2018), 10.1080/15583058.2018.1492647
6. S. Galassi, N. Ruggieri, G. Tempesta, Ruins and archaeological artifacts: vulnerabilities analysis for their conservation through the original computer program BrickWORK, in: Aguilar, R., Torrealva, D., Moreira, S., Pando, M., Ramos, L.F. (Eds.), *Structural Analysis of Historical Constructions*, RILEM bookseries 18, Springer International Publishing, pp. 1839-1848 (2018) 10.1007/978-3-319-99441-3_197, Proc. of 11th International Conference on structural analysis of historical constructions (SAHC2018)(11-13 September 2018, Cusco, Perù)
7. S. Galassi, G. Misseri, L. Rovero, G. Tempesta, Failure modes prediction of masonry voussoir arches on moving supports, Eng. Struct. **173**, 706-717 (2018) 10.1016/j.engstruct.2018.07.015
8. S. Galassi, G. Misseri, L. Rovero, G. Tempesta, Equilibrium analysis of masonry domes. On the analytical interpretation of the Eddy-Lévy graphical method, Int. J. Archit. Herit. **11**, 8, 1195-1211 (2017) 10.1080/15583058.2017.1372823
9. F. Pugi, S. Galassi, 2013 Seismic analysis of masonry voussoir arches according to the Italian building code, Ing. Sismica-Ital. **30** ,3, 33-55 (2013)
10. M. Paradiso, S. Galassi, S. Benedetti, Stabilità di archi e volte in muratura: metodi tradizionali e calcolo automatico, in: Luis Villegas, Ignacio lombillo, Clara Liano, HaydeeBlanco (Eds.), *Patologia della Costruccion, Tecnologia de la Rehabilitacion y Gestion del Patrimonio*, Proc. Conf Latinoamericano Patologie del costruito, tecniche di riabilitazione e gestione del patrimonio (Rehabend 2014), Santander, Spain, Gráficas Iguña, S.A. (2014)
11. S. Galassi, M. Paradiso, BrickWORK Software-Aided Analysis of Masonry Structures, IERI Procedia7, 62-70 (2014), Special Issue 2013 International Conference on Applied Computing, Computer Science, and Computer Engineering (ICACC 2013), USA, December 2013, Elsevier B.V., ISSN: 2212-6678, 10.1016/j.ieri.2014.08.011
12. M. Paradiso, S. Galassi, A. Borri, D. Sinicropi, "Reticolatus": an innovative reinforcement for irregular masonry. A numeric model, in: Paulo J. S. Cruz (Ed.), *Structures and Architecture: Concepts, Applications and Challenges*, CRC Press / Balkema, Taylor and Francis Group, London, Proc. 2nd Int. Conf. on Structure & Architecture ICSA2013, Guimaraes, Portugal (2013)
13. S. Galassi, A numerical procedure for failure mode detection of masonry arches reinforced with fiber reinforced polymeric materials, IOP Conf. Ser.: Mater. Sci. Eng. **369**, 012038 (2018) 10.1088/1757-899X/369/1/012038, Proc. of 5th Global Conference on Polymer and Composite Materials (PCM 2018) (10-13 April 2018, Kitakyushu City, Japan)
14. S. Galassi, Analysis of masonry arches reinforced with FRP sheets: experimental results and numerical evaluations, MATEC Web of Conferences207: 01002, 10.1051/matecconf/201820701002, Proc. of International Conference on Metal Material Processes and Manufacturing (ICMMPM2018) (19-20 July 2018, Jeju Island, South Korea)
15. D. Sinicropi, E. Perria, S. Galassi, M. Paradiso, A. Borri, Artificial Ageing of Mortar Prisms Reinforced through Steel, Glass and Organic Fibers, Journal of Key Eng. Mat., A. Di Tommaso, C. Gentilini and G. Castellazzi (Eds.) **624**, 4, 542-50 (2014) 10.4028/www.scientific.net/KEM.624.542
16. P.G. Hatcher, Wood associated with the A.D. 79 eruption, its chemical characterization by solid state as a guide to the degree of carbonization. In: W.F., Jashemski, F.G., Meyer (Eds.), The Natural History of Pompeii, Cambridge University Press, Cambridge (2002)
17. M.F. Sheridan, F.Barberi, M.Rosi, R.Santacroce, A model for Plinian eruptions of Vesuvius. Nature, **289**, 282-285 (1981)
18. A.Ciarallo, 2012, Gli spazi verdi dell'antica Pompei, Aracne Editore, Roma.
19. G. Di Pasquale, E.Allevato, A.Cocchiararo, D. Moser, M.Pacciarelli, A.Saracino, 2014, Late Holocene persistence of Abies alba in low-mid altitude deciduous forests of central and southern Italy: new perspectives from charcoal data, J. Vegetation Science (2014) 10.1111/jvs.12196
20. A. Ciarallo, *Flora pompeiana* (L'Erma di Bretschneider, Roma, 2004)
21. D. Camardo, M.Notomista, The Roof and the Suspended Ceiling of the Marble Room in the House of Telephus Relief at Herculaneum, J. Roman Archaeology, **28**, 39-70 (2015)
22. P. Kastenmeier, D. Camardo, M. B. Casieri, A. K.-U. D'Andrea, Heussner, M. Notomista, Studio delle tracce di lavorazione, dendrocronologia e documentazione sui legni del tetto della Casa del

Rilievo di Telefo ad Ercolano, Bullettino dell'Istituto Archeologico Germanico, Sezione Romana, 121, 269-310 (2015)
23. M. V. Antico Gallina, *Archeologia del legno* EDUCatt Università Cattolica, Milano, 2011
24. N. Ruggieri, Carpenteria di legno dei tetti e dei solai interpiano a Pompei nel I secolo d.C., Restauro Archeologico, ISSN 1724-9686 (print) ISSN 2465-2377 (online), **2**, 4-19 (2017), Firenze University Press10.13128/RA-22204

Antifungal Property of *Piper betle* Leaf Oil against Oral Candida Species

Rattiporn Kaypetch[1] and Sroisiri Thaweboon [2a]

[1]*Research Office, Faculty of Dentistry, Mahidol University, Bangkok, Thailand*
[2]*Department of Oral Microbiology, Faculty of Dentistry, Mahidol University, Bangkok, Thailand*

Abstract. Fungal infection is one of the main clinical problems due to the extensive uses of broad-spectrum antibiotics and immunosuppressive therapy. Among all, candida species are the most prevalent. *Piper betle* Linn., a tropical plant intimately associated with pepper, has been widely used as a traditional herb in many Asian countries. The purpose of this study was to evaluate the antimicrobial effect of essential oil extracted from fresh leaves of *P. betle* against four strains of candida species, *C. albicans, C. glabrata, C. krusei, C. parapsilosis, C. tropicalis, C. pseudotropicalis* and *C. stellatoidia*. Inhibitory activity was primarily screened by Kirby-Bauer disc diffusion technique and subsequently the minimum inhibitory concentration (MIC) was determined by agar dilution technique. Betel oil exhibited a high potential of antifungal property against all strains of yeast with inhibition zones ranged from 32 to 33 mm. in diameter and MIC values of 0.039-0.078 % v/v. Data from this study demonstrates a potential application of betel oil in drug preparations and development for the treatment of candida infection. Further investigations are required to define the antifungal mechanism of this oil as well as clinical trial in the patients.

1 Introduction

Recently, there has been a rising interest in opportunistic infections caused by pathogenic fungi. The growing concern is due to an increasing population with HIV infection and the more extensive uses of broad-spectrum antibiotics, immunosuppressive therapy and invasive surgical procedures. Of the fungi regarded as the cause of infection, candida species are the most commonly found. The genus candida comprises of a heterogeneous group of yeasts with at least 50 different species. Nonetheless, approximately 70% of invasive infections are caused by *Candida albicans, Candida glabrata, Candida, parapsilosis, Candida tropicalis, C. pseudotropicalis* and *Candida krusei* [1]. The virulence of these yeasts is ascribed to the capacity to avoid host defenses, adherence and biofilm formation on host tissue, and the releasing of hydrolytic enzymes that can damage tissues [2].

In dentistry, candida infection in the oral cavity, oral candidiasis, can present in many forms according to how deeply the organism has spread, or if host defenses allow for more consequential infections. The most common form of candida infection is oral thrush which is also known as pseudomembranous candidiasis. This form of mucosal infection is usually described by a white cottage cheese–like plaque or biofilm that can be wiped off to disclose a red bleeding base. The plaque that can be removed is generally composed of aggregates of the candida hyphae and by products of epithelial breakdown. This infection is seen in high risk people such as infants and young children, whose host defenses have yet to develop, or in the elderly with a decline in immune defenses.

In addition to infants and aging people, oral candidiasis has increased due to the spread of HIV infection. In this regards, oropharyngeal candidiasis is a clinical indicator of HIV disease advancement.

Few antifungal drugs are accessible compared with the wide range of antibiotics for bacterial infections. This may indicate the difficulty implicated in developing an agent with property against a eukaryotic cell type without host toxicity. Some commercially available antifungal agents are brought to use for the treatment of candida infection, however, these drugs had been reported to produce adverse effects such as allergic reactions, interaction with other medicines, and drug resistance.

In order to solve the unfavourable effects of commercial products and to control candida infection, the development of natural products capable of clinical application is required. Natural products have been evidenced to be an alternative to synthetic chemical agents and the remark in medicinal herbs as a source of antimicrobial agents has increased rapidly. *Piper betle* Linn. or betel is a tropical plant intimately associated with pepper which is extensively grown in many Southeast Asian countries such as Thailand, Malaysia and India. It is a root climber with deep green heart shaped leaves. Betel leaves are the most important part and are of medicinal, religious and ceremonial value. Chewing betel leaves is supposed to prevent bad breath, harden the gum, conserves the teeth and sweetens breath. The infusion prepared from the leaves has been known as traditional herbs used as mouthwash, cough medicine,

[a] Corresponding author: sroisiri.tha@mahidol.ac.th

tonic or astringent [3]. Recently, oil from betel leaves has been concluded to be a promising agent in controlled and released food packaging technology by its antimicrobial activity against food spoilage yeasts [4].

The purpose of this study was to evaluate the antifungal effect of essential oil extracted from fresh leaves of *P. betle* against seven strains of candida species.

2 Materials and Methods

2.1 Oil

Betel oil was extracted from fresh leaves of *P. betle* by hydrositillation by the Thailand Institute of Scientific and Technological Research (TISTR) and prepared in 5% dimethyl sulfoxide (DMSO).

2.2 Candida species

The yeast strains used in the study were *Candida albicans* ATCC 10231, *Candida krusei* ATCC 6258, *Candida glabrata* ATCC 90030, *Candida parapsilosis* ATCC 22019, *Candida tropicalis* (clinical isolate), *Candida pseudotropicalis* (clinical isolate) and *Candida stellatoidia* (clinical isolate). All of them were obtained from a culture collection of Oral Microbiology Department, Faculty of Dentistry, Mahidol University, Thailand. Sabouraud dextrose broth (Difco, USA) was used to culture all yeast strains and the suspension of 10^7 CFU/ml of yeast (equivalent to 0.5 McFarland Standard) was used for antimicrobial evaluation.

2.3 Disc diffusion test

The inhibitory potential of betel oil on fungal growth was evaluated using the disc diffusion technique described by Kirby-Bauer [5]. In brief, freshly prepared yeast suspension was spread on Mueller Hinton agar (Difco, USA). A 20 μL of betel oil was dropped onto sterile paper disc (6 mm. in diameter) and left for 60 sec. Then, paper disc soaked with oil was put on the agar surface and the plate was incubated at 37°C for 24-48 h. The same volume of 0.2% chlorhexidine gluconate solution and 5% DMSO solution served as positive and negative controls, respectively. After incubation, the zones of inhibition surrounding paper disc were measured using a Vernier caliper.

2.4 Minimum inhibitory concentration (MIC)

The MICs of betel oil on candida were investigated with the agar dilution technique. Mueller Hinton agar plates were prepared with serial two-fold dilutions of betel oil (ranged from 20 to 0.02 % v/v). The yeast suspensions were prepared by adjusting their density to 0.5 McFarland Standard. The 20 μL of suspension were dropped on each plate and incubated at 37 °C for 48 h.

All tests were carried out in triplicate on three consecutive times.

3 Result

By using disc diffusion method, betel oil exhibited inhibitory effect on the growth of candida species (Table 1). The minimum inhibitory concentration (MIC) ranged from 0.039 to 0.078 % v/v (Table 2). According to higher MIC value, *C. albicans, C. krusei* and *C. tropicalis* were more resistant to the oil.

Table 1. Inhibitory effect of betel oil against various candida species

Candida species	Inhibition zone (mm)		
	oil	+ve control[a]	-ve control[b]
C. albicans	33.83±0.76	22.83±0.28	-
C. glabrata	33.83±0.76	23.50±0.50	-
C. krusei	32.66±0.57	23.33±0.57	-
C. parapsilosis	33.83±0.76	24.33±0.57	-
C. tropicalis	30.83±0.28	23.83±0.28	-
C. pseudotropicalis	33.50±0.50	24.16±0.28	-
C. stellatoidia	35.50±0.86	25.83±0.28	-

[a] positive control was 0.2% chlorhexidine gluconate
[b] negative control was 5% DMSO
values are expressed as mean ± SD

Table 2. Minimum inhibitory concentration (MIC) values of betel oil against various candida species.

Candida species	MIC (% v/v)
C. albicans	0.078
C. glabrata	0.039
C. krusei	0.078
C. parapsilosis	0.039
C. tropicalis	0.078
C. pseudotropicalis	0.039
C. stellatoidia	0.039

4 Discussion

Regarding the rising occurrence of drug resistant pathogenic fungi and the toxicity of current antifungal drugs, the search for natural products that exhibit antimicrobial properties as an alternative treatment has been suggested. Plants have been used in medicine for a long time since they represent an economic alternative, are easily obtainable and can be applied to several pathologies [6]. *Piper betle* leaves have been traditionally used in folk medicine for the cure of various disorders and for chewing purposes together with other seasoning. The ethnic groups in many Asian countries such as Sri Lanka, India, and Thailand, chew the leaves as a narcotic which induces swoon and excessive sweating and also helps to give warmth to the body during the cool season. Betel oil has been shown to possess antimicrobial, gastroprotective, hepatoprotective, wound healing and antioxidant activities [7].

The results from the present study demonstrated antifungal properties of betel oil against all tested strains of candida with a zone of inhibition ranged from 32 to 35 mm. in diameter compared with 22 to 25 mm. of chlorhexidine solution, a positive control. The MIC values of the oil against candida were 0.039-0.078 % v/v. In this regards, *C. albicans, C. krusei* and *C. tropicalis* exhibited more resistant to the oil. Previous studies have been conducted using essential oil and ethanolic extract of *P. betle* to evaluate antimicrobial activity against bacteria and fungi [3,7]. The growth inhibitory effect was reported against various gram positive bacteria (*Enterococcus faecalis, Streptococcus pyogenes, Bacillus cereus, Micrococcus luteus*), gram negative bacteria (*Escherichia coli, Salmonella enteritidis*) and yeasts (*Candida albicans, Saccharomyces cerevisiae*). In the field of dentistry, anti-halitosis activity of betel leaf extract was demonstrated due to the reduction of methyl mercaptan and hydrogen sulfide produced by obligate oral anaerobes (*Porphyromonas gingivalis* and *Fusobacterium nucleatum*) [8]. Furthermore, it has also been reported that aqueous extract could inhibit growth, adherence and extracellular polysaccharide formation of caries-associated bacteria (*Streptococcus mutans, Streptococcus mitis* and *Actinomyces* spp.) suggesting its potential role in controlling dental plaque development [9,10].

P. betle contains a vast variety of biologically active components whose concentration bases on the diversity of the plant, climate, harvesting period and geographic location. The major components of betel oil were reported to be safrole, chavibetol acetate, allylpyrocatechol diacetate, methyl chavicol, methyl eugenol, 1,8-cineole, eugenol, caryophyllene, carvacrol, pinene, limonene, campene and cadinene [7,11]. However, the most active antimicrobial components appear to be carvacrol, eugenol and chavibetol [12]. The antifungal effects were attributed to their ability to inhibit ergosterol biosynthesis of yeast cell and to make the cytoplasmic membrane porous [13]. These result in the disruption of cytoplasmic membrane integrity. In addition, active component such as carvacrol has a hydroxyl group on the phenolic ring of its structure which increases hydrophilic ability and help it dissolve in the microbial membranes and impair them. Previous study [14] proposed mechanism of action of *P. betle* extract on candida that it involved the aberration of cell wall elements and the damage of cytoplasmic membrane, causing coagulation of cell components and the efflux of critical molecules and ions which lead to an autolytic phase that finally brings about cell death.

In the case of safety profile, studies in animal models revealed that betel oil was safe in terms of hepatotoxicity, renotoxicity, hematotoxicity, gross morphology, stress or aversive behaviours [15,16]. Oral administration of betel extract showed protection against gastric damage induced by absolute ethanol by the increase of mucus content [17]. Furthermore, Nilugal *et al* [18] reported that ointment containing betel extract could enhance the rate of wound contraction and reduce the healing period when applied to rats.

In conclusion, data from this in vitro study demonstrate that essential oil from *P. betle* leaves exhibits inhibitory effect on candida spp. It might have a potential application in drug preparations and development for the treatment of candida infection. However, further investigations are required to define the antifungal mechanism of this oil as well as clinical trial in the patients.

Acknowledgement

The Authors would like to thank the Thailand Institute of Scientific and Technological Research (TISTR) for providing us with *Piper betle* oil, along with Associate Prof. Boonyanit Thaweboon for his valuable advice on various technical issues, and Miss Thaniya Muadcheingka for her kind assistance in preparing culture media.

References

1. M.A. Pfaller, Diekema, Clin. Microbiol. Rev. **20**, 133 (2007)
2. S. Silva, M. Negri, M. Henriques, R. Oliveira, D.W. Williams, J. Azeredo, F.E.M.S. Microbiol. Rev. **36**, 288 (2012)
3. M.P. Rai, K.R. Thilakchand, P.L. Palatty, P. Rao, S. Rao, H.P. Bhat, M.S. Baliga, Asain. Pacific. J. Cancer. Prev. **12**, 2149 (2011)
4. P. Suppakul, N. Sala-Ead, P. Phoopuritham, Kasetsart. J. **40**, 91 (2006)
5. A.W. Bauer, W.M. Kirby, J.C. Sherris, Am. J. Clin. Pathol. **36**, 493 (1966)
6. M.C. Cruz, P.O. Santos, A.M. Jr Barbosa, D.L. de Mélo, C.S. Alviano, A.R. Antoniolli, D.S. Alviano, R.C. Trindade, J. Ethnopharmacol. **111**, 409 (2007)
7. L.S. Arambewela, L.D. Arawwawala, K.G. Kumaratungan, D.S. Dissanayake, W.D. Ratnasooriya, S.P. Kumarasingha, Pharmacogn. Rev. **5**, 159 (2011)
8. N. Ramji, R. Iyer, S. Chandrasekaran, J. Ethnopharmacol. **83**, 149 (2002)
9. Z.H. Rahim, N. Thurairajah, J. Appl. Oral. Sci. **19**, 37 (2011)
10. F.A. Razak, Z.H. Rahim, J. Oral. Sci. **45**, 201 (2003)
11. A.M. Rimando, Arch. Pharm. Res. **9**, 93 (1986)
12. H.J. Dorman, S.G. Deans, J. Appl. Microbiol. **88**,

https://doi.org/10.1051/matecconf/201824201021

308 (2000)
13. A. Ahmad, A. Khan, F. Akhtar, S. Yousuf, I. Xess, L.A. Khan, N. Manzoor, Eur. J. Clin. Microbiol. Infect. Dis. **30**, 41 (2011)
14. M.A. Nordin, W.H. Wan Harun, F. Abdul Razak, M.Y. Musa, Int. J. Oral. Sci. **6**, 15 (2014)
15. L.S. Arambewela, W.D. Ratnasooriya, L.D. Arawwawala, Trop. Med. Plants. **4**, 195 (2003)
16. A. Sengupta, P. Adhikary, B.K. Basak, K. Chakrabarti, P. Gangopadhyay, J. Banerji, A. Chatterjee, Indian. J. Exp. Biol. **38**, 338 (2000)
17. L.D. Arawwawala, L.S. Arambewela, W. Ratnasooriya, J. Ayurveda. Integr. Med. **5,** 38 (2014)
18. K.C. Nilugal, K. Perumal, R.E. Ugander, Am. J. Pharm. Tech. Res. **4,** 443 (2014)

Finite element simulation and microstructure analysis of nickel based alloy for additive manufacturing

Liu Heping [12,a], Sun Fenger [1], Gao Yibo[1], Cheng Shaolei[1], and Liu Bin[1]

[1] *School of Materials Science and Engineering, North University of China, Taiyuan 030051, China*
[2] *School of Materials Science and Engineering, Taiyuan University of Technology, Taiyuan 030024, China*

Abstract. In this paper, the finite element simulation of GH4169 high temperature alloy by selective laser melting was carried out, and the microstructure was analyzed by experiments. The results show that the shape of the temperature field cloud formed by the laser heat source is different from the shape of the theoretical model, but is in the shape of the ellipse. The temperature gradient at the front end of the molten pool is larger than that of the back end of the molten pool, and the isotherm of the front end of the molten pool is more intensive. The temperature of the substrate is less affected by the temperature gradient. The temperature gradient of the front end of the melting pool is larger than the back end of the molten pool, and the temperature field of selective laser melting is like a meteor with trailing tail. In the laser heat source, the temperature isotherm is the most dense and the temperature gradient is maximum. The relative effect of mechanical properties of δ phase is very complex. When the phase is precipitated by widmanstatten structure, it is easy to produce stress concentration as a source of cracks

1 Introduction

Ni based superalloys have high oxidation resistance, high temperature resistance, hot corrosion resistance and fatigue resistance. It has been widely used in anti-aircraft engine components, such as turbine blades and combustors, which work for a long time in high temperature environment. For the manufacture of superalloys, the methods of casting, powder metallurgy and precision machining are mainly used at present[1-5]. The cost is high and the cycle is long in the process of part design and die manufacturing. In order to further meet the requirements of material properties in aviation field, alloying elements with different contents are often added into materials. This will make the material composition more and more complex and the hot processing performance will be worse. Moreover, traditional casting methods can cause severe component segregation, which will lead to unstable microstructure and properties. With higher requirements for service performance, powder metallurgy technology can be gradually applied to aeroengine turbine blades, turbine discs and other high temperature parts. The parts formed by powder metallurgy have fine grains and uniform structure, but there are some defects such as heterogeneous inclusions, thermal induced voids and original grain boundary. Therefore, materials researchers have been studying that the only way to develop superalloys is to combine them with additive manufacturing technology[6-9].

Additive manufacturing technology combines computer-aided design, material processing and forming technology. Based on digital model files, special metal materials, non-metal materials and medical biomaterials are stacked layer by layer through software and numerical control system to manufacture physical objects. In recent years, this technology has become the mainstream direction of manufacturing technology development because of its advantages of fast research and development speed, low cost risk and high flexibility.

In this paper, GH4169 alloy powder is taken as the research object, and the effect of technological parameters on the formability of nickel-based alloy material is analyzed.

2 Experiment and simulation method

2.1 Experimental method

Nickel based alloy powder GH4169 was used as melting material with 250 mesh. Ball milling parameters are ball milling ratio 10:1, speed 140r/min and ball milling 2h. The laser wavelength is 1.06 um, the minimum spot diameter of laser beam 0.15 mm, and the thickness of metal powder 0.05-0.30 mm.

2.2 Simulation method

In the transient temperature field, the temperature is not only related to the three-dimensional change of the space, but also a function of time. In the two-dimensional

[a] Corresponding author: peace666@126.com

transient temperature field, the finite element method usually divides the three-dimensional model into several finite elements to solve the instantaneous temperature distribution. This method improves the efficiency of temperature field analysis. Using the mathematical calculation model in the selected cell, the node temperature T_i is the discrete unit[10-13].

$$T = T = \sum_{i=1}^{n_e} N_i(x,y)T_i(t) \quad (1)$$

where, T is a node temperature array and a reciprocal array for node temperature to time.

The heat conduction model of each unit is calculated as follows:

$$K_{ij}^e = \int_{\Omega^e} \left(k_x \frac{\partial N_i}{\partial x}\frac{\partial N_j}{\partial x} + k_y \frac{\partial N_i}{\partial y}\frac{\partial N_j}{\partial y} \right) d\Omega \quad (2)$$

The model calculation formula of heat transfer boundary condition of unit body can be expressed as:

$$H_{ij}^e = \int_{\Theta_3^e} h N_i N_j d\Theta \quad (3)$$

The specific heat capacity model of the unit is:

$$C_{ij}^e = \int_{\Omega^e} \rho c N_i N_j d\Omega \quad (4)$$

In the temperature field simulation, the temperature of the element node is the instantaneous temperature at that time, that is:

$$T_i = T_i \quad (i = 1,2,3,\ldots n) \quad (5)$$

3 Results and discussion

3.1 Temperature field analysis of single channel scanning process

In the process of laser melting experiment, single-pass scanning is the basic prerequisite for optimizing process parameters. The quality of scanning line will affect the quality and performance of the whole forming workpiece. In order to reduce the number of tests and save costs, the finite element simulation method is proposed to be used to effectively select process parameters.

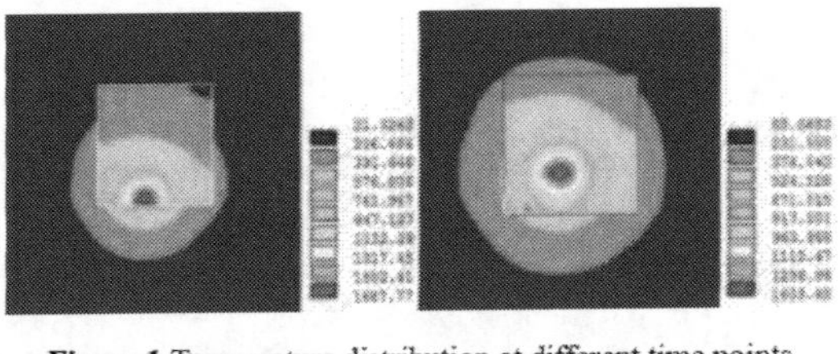

Figure 1 Temperature distribution at different time points during single channel scanning (a) 0.015s (b) 1.1s

Figure 1 (a) shows the first step of 0.015s single-pass scanning, the temperature distribution cloud of alloy powder at the beginning of melting. As can be seen from the figure, the maximum temperature of the spot center is 1687.77 ℃, reaching the melting point of the alloy powder, and the powder can be melted. Figure 1 (b) is a cloud map of temperature distribution at 1.1s. It can be seen that the maximum temperature of the spot center dropped to 1403.45℃ at this time. From the temperature field distribution, it can be clearly concluded that the substrate is less affected by the heat, and the basic temperature is too high to affect the powder bed. The temperature isotherm of the rear end of the molten pool is sparse, and the temperature isotherm of the front of the pool is dense. The pool temperature and the size of the molten pool gradually increase with the increase of absorbed heat.

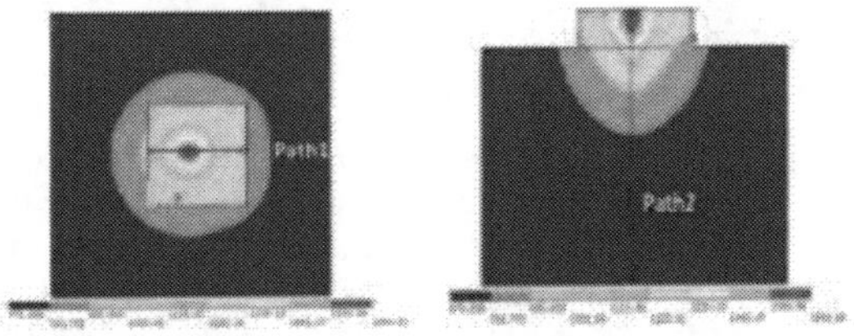

Figure 2 The temperature field distribution in single channel scanning at 0.15s (a) 10th substep (b) Cross section along the X axis

Figure 2(a) is a temperature field distribution cloud diagram from the single pass scanning to the 10th substep 0.15s. Figure 2 (b) is a temperature field distribution diagram of the molten pool along the X-axis transverse section at that time. As can be seen from the figure, the temperature gradient at the front end of the molten pool is larger than the temperature gradient at the rear end of the molten pool. Moreover, the isotherms at the front end of the molten pool are denser. The main reason for this phenomenon is that the maximum temperature reaches 1,664.84°C due to the action of the heat source. The temperature exceeds the melting point of the alloy powder. The powder area is now completely melted. However, the substrate and ambient temperature are low and forced cooling occurs. Therefore, a large temperature gradient is generated As the heat is transferred, the temperature of the substrate and the surrounding environment will gradually increase. In turn, the heat affected zone is gradually enlarged.

3.2 Temperature distribution at the end of each scan

Figure 3 is an isotherm diagram at the end of the melting and sintering of each sintering line in a single layer scan. As can be seen from the figure, the maximum temperature at the end of the first scan line can reach 1829.26 °C. At the end of the second scan line, the maximum temperature is 1524.73 °C. The maximum temperature at the end of the third scan line is 1494.63 °C. At the end of the fourth scan line, the maximum temperature is 1602.96 °C. At the end of the fifth scan line, the maximum temperature is 1608.18 °C. The maximum temperature at the end of the sixth scan line is 1988.45 °C. The heat generated by the laser heat source during the first pass of sintering has a greater effect on the powder. However, the influence on the substrate is small. As the laser heat source moves gradually, the scan line will gradually increase. The accumulation of heat causes the temperature of the powder bed to rise. Further, the temperature of the substrate is further increased. The temperature change is generally stable throughout the single layer scan. There were no high or low temperatures during the period. Because the latent heat of phase change of powder melting is considered in the simulation design. This avoids the heat accumulation effect causing the substrate temperature to be too high to affect the formation of the alloy powder. From the above analysis, it can be seen that under the action of the laser heat source. The laser heat source has a certain preheating function for the area around the powder bed. However, the thermal impact on the substrate is not very large. At the same time, the temperature isotherm at the action of the laser heat source is the most dense, that is, the temperature gradient is the largest.

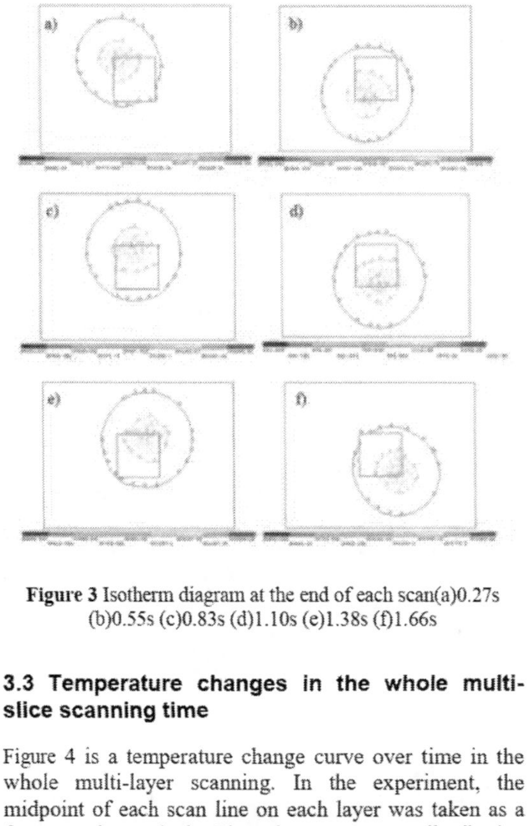

Figure 3 Isotherm diagram at the end of each scan(a)0.27s (b)0.55s (c)0.83s (d)1.10s (e)1.38s (f)1.66s

3.3 Temperature changes in the whole multi-slice scanning time

Figure 4 is a temperature change curve over time in the whole multi-layer scanning. In the experiment, the midpoint of each scan line on each layer was taken as a feature point analysis. First, the temperature distribution is divided into three stages. Each stage corresponds to the scan time period of each layer. The temperature trends of the six points in each layer are similar and cyclical. At the same time, the number of peaks per layer is the same. The characteristic point is that the maximum temperature is gradually increased. This is mainly due to the effect of heat accumulation during the forming process. However, the temperature of the entire block forming process is relatively stable. This ensures that the performance of each layer remains substantially the same during the multi-layer scanning process.

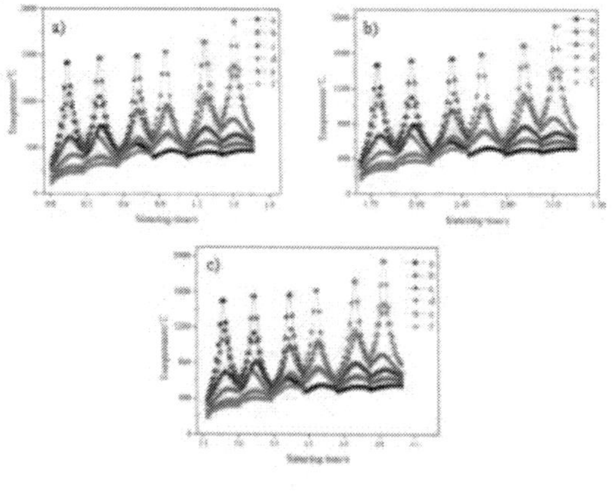

Figure 4 The temperature-time history curve in multi layer scanning(V=150mm/min,P=150W,Ψ=33%)

3.4 Microstructure analysis

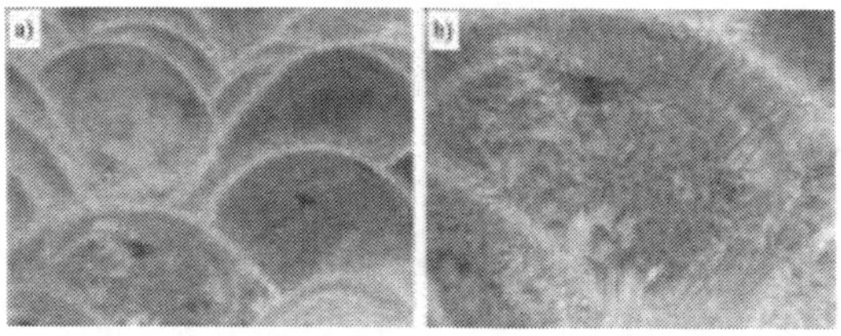

Figure 5 The morphology of nickel based alloy (a)Macroscopic morphology,(b)Local amplification

Figure 5 shows the weld pool morphology of Ni based alloy. In the experiment, the thickness of powder laying is 0.2 mm, the width of molten pool measured 0.072-0.12 mm, and the average thickness 0.065-0.10 mm. The results show that the overlap degree between adjacent molten pools greatly affects the density of the formed parts. The density of formed specimens will also increase with the decrease of scanning spacing, but too small scanning spacing will cause warping and cracking of formed parts.

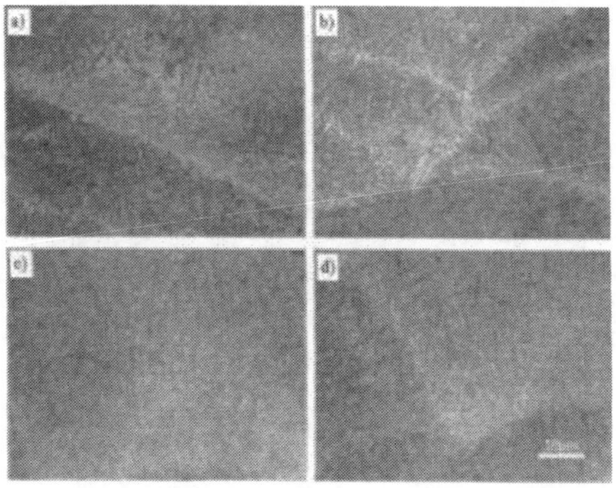

Figure 6 Microstructure of different parts of forming parts

Figure 6 is the microstructure of different parts of the melding parts. The lighter colour is where the cladding layers overlap each other, and the area of the structure is similar to the raft structure of single crystal nickel-based alloys. The microstructure of each position is very compact. In general, GH4169 alloy precipitates δ phase, γ″phase and γ′ phase after solid solution and aging heat treatment[14,15]. Among them, the γ″and γ′phases maintain a coherent relationship with the matrix, mainly through the coherent strain to strengthen the alloy. The high volume fraction of the γ″ phase is the first condition to ensure high strength of the alloy. The effect of δ phase on the relative mechanical properties is very complicated. When the δ phase is precipitated in a large amount in the Wei's structure, it is easy to cause stress concentration to become a crack source. At the same time, the content of the substrate Nb is reduced. The γ" phase cannot be precipitated around it, and the strength of the alloy is lowered. However, when a proper amount of delta phase distributes along grain boundaries, the stress state at the front of the crack will be relaxed. The formation and propagation of the crack will be delayed, and the durability of the crack will be improved.

4 Conclusion

Laser power and scanning speed are the main parameters that affect temperature changes. The size of the bath is proportional to the instantaneous maximum temperature and laser power. But it is inversely proportional to the scanning speed. The shape of the temperature field cloud formed by the action of the laser heat source is different from the shape of the theoretical model, but is in an elliptical state. The temperature gradient at the front end of the bath is greater than the temperature gradient at the back end of the bath. Moreover, the isotherms at the front end of the molten pool are denser.

The substrate is less affected by the temperature. A large temperature gradient is formed at the laser scanning. The temperature gradient at the front of the bath is greater than the back end of the bath. The laser forming temperature field distribution is like a meteor with a tail. The temperature isotherm at the action of the laser heat source is the densest, that is, the temperature gradient is the largest. The characteristic point is that the maximum temperature is gradually increased. This is mainly due to the effect of heat accumulation during the forming process. The temperature variation over the entire multi-layer scan time history is characterized by a gradual increase in the maximum temperature. This is mainly due to the effect of heat accumulation during the forming process.

The degree of overlap between adjacent molten pools greatly affects the density of the molded parts. The density of the molded part also gradually increases as the scanning pitch decreases. However, too small scanning distance will cause warping and cracking of molding parts.

Acknowledgments

This research work was financially funded by the Science and Technology Innovation Project of Shanxi Province (No.2016156), the Postdoctoral Science Foundation of China (No.2016M590214), and the Natural Science Foundation of Shanxi Province, China (No. 2015011036).

References

1. Ruiz-Sabariego J A, Pommier S. International Journal of Fatigue, 31,11 (2017)
2. Rai R K, Sahu J K. Materials Letters, 220.(2018)
3. Hosseini S A, Abbasi S M, Madar K Z, et al. Materials Chemistry & Physics, , 211 (2018)
4. Lin Y C, Nong F Q, Chen X M, et al. Vacuum, 137(2017)
5. Shi X, Jin X, Lin H, et al. Journal of Alloys & Compounds, 693(2017)
6. Gibson I, Rosen D W, Stucker B. Springer US, (2010)
7. Kruth J P, Leu M C, Nakagawa T. CIRP Annals - Manufacturing Technology, 47,2(1998)
8. Li W, Yang K, Yin S, et al. Journal of Materials Science & Technology, 3(2018)
9. Song J, Chew Y, Bi G, et al. Materials & Design, 137(2018)
10. Hama T, Nagata T, Teodosiu C, et al. International Journal of Mechanical Sciences, 50,2 (2008)
11. Hussein A, Hao L, Yan C, et al. Materials & Design, 52,24 (2013)
12. Gan Y , Wang W , Cui Z , et al. Optik - International Journal for Light and Electron Optics, 126,(7-8) (2015)
13. Singh S K, Mahesh K, Kumar A, et al. Materials & Design, 31,9 (2010)
14. Li Z, Tian S, Zhao Z, et al. Journal of Aeronautical Materials, 31,5 (2011)
15. Zhang X C, Li H C, Zeng X, et al. Materials Science & Engineering A, 682 (2017)

AUTHOR INDEX

EDP Sciences – Web of Conferences
17, Avenue du Hoggar
Parc d'Activité de Courtabouf
BP 112
F-91944 Les Ulis Cedex A
France

ISBN 978-1-5108-7596-8

9 781510 875968

CURRAN ASSOCIATES INC.
proceedings
.com

International Conference on Materials Applications and Engineering (ICMAE 2017)

MATEC Web of Conferences Volume 142 (2018)

Qingdao, China
25 - 27 August 2017

edp sciences